P9-DDI-100

Base Instincts

Base Instincts

What Makes Killers Kill?

Jonathan H. Pincus, M.D.

W. W. NORTON & COMPANY
New York • London

Printed in the United States of America
First Edition

The text of this book is composed in Sabon with the display set in Sabon Bold
Composition by Sue Carlson
Manufacturing by Haddon Craftsmen, Inc.
Book design by Charlotte Staub
Production manager: Julia Druskin

Library of Congress Cataloging-in-Publication Data

Pincus, Jonathan H., 1935–
Base instincts : what makes killers kill? / by Jonathan H. Pincus.
p. cm.
Includes bibliographical references.
ISBN 0-393-05022-X
1. Murder—Psychological aspects. 2. Murders—Psychology. I. Title.
HV6515 .P557 2001
364.15'23'019-dc21 00-069572

W. W. Norton & Company, Inc., 500 Fifth Avenue, New York, N.Y. 10110
www.wwnorton.com

W. W. Norton & Company Ltd., Castle House, 75/76 Wells Street, London W1T 3QT

1 2 3 4 5 6 7 8 9 0

Contents

Acknowledgments

Without my dear friend, Dorothy Otnow Lewis, M.D., there would have been no book to write. It was she who cajoled, goaded, and convinced me to participate with her in our earliest studies of violent criminals, and she has guided me ever since. At the beginning, she had a tough job. As a "regular" neurologist, I shared the prejudices of my neurological colleagues toward disorders of cognition and behavior. The only real diseases of the brain, according to that paradigm, were disorders of sensation, movement, and speech. Dorothy insisted that behavior in general and violence in particular originate in the brain, that the subject of violence was within my purview as a neurologist. She was right. I owe her an enormous debt of gratitude for her inspiration and encouragement.

There are many others whose wisdom, assistance, and friendship provided crucial support to me in the writing of this book. First among these are Joyce Cowan and Emily Wilson. They edited and criticized constructively and sensitively. I must take full responsibility for any failings of this book but they share importantly in its good features.

Others who provided crucial help are Hilary Sheard, Jane Lipton, Jean Holz, Elliott Wilner, Leon Wieseltier, Majorie Sexton, Lee Ann Konstantinov, Barbara Harslem, Jutta Lewis, Pamela Blake, Kathy Schneider, T. Rae Van Dyk, Vivian Valenzuela, Sean O'Brien, and, of course, my agent, Gail Ross, and my editor at W. W. Norton, Angela von der Lippe.

Prologue

A fifteen-year-old boy named Kip Kinkel from Springfield, Oregon, shot and killed his parents. The next day, he went to school and opened fire at schoolmates and teachers alike. Newspaper accounts provided diagrams of the school, the location of each victim's body, and the path of destruction. In all this detail, there was not a clue as to what made this young man kill. I wondered from what kind of home he came, from what mental illness he suffered, and whether he was neurologically abnormal. When I examined him, he told me of voices that directed him to kill. He described the gun-drenched culture of his family, the verbal abusiveness of his parents, and his sense of loneliness and being threatened. His physical examination was abnormal and imaging tests of his brain showed abnormalities as well.

Not long after Kinkel's crime, Russell Weston entered the U.S. Capitol in Washington, D.C., and shot and killed two guards before being severely wounded and subdued. What compelled him to do it? He had been diagnosed as mentally ill and had been involuntarily hospitalized in Montana before the killing. The diagnosis in Montana had been paranoid schizophrenia In the hospital in which he was recovering from his wounds, he told me of the "Ruby Satellite Project" that he had to forestall. He was completely mad, but there are many paranoid schizophrenics, equally mad, who have never killed anyone. Why had this particular man become a killer? From what sort of home did he come? The news reports provided every detail concerning what he did but very little about the internal and external forces that pressed him to kill.

These killings can be classified as mass murders as more than one person was killed, and the killer had no personal grudge against most of the people he killed. Investigating one hundred mass murderers, the *New York Times* recently reported that more than half were individuals with a history of severe psychopathology that preceded the crime. I believe that the cause of such horrors was closely related to the motivation of Susan Smith, who strapped her two toddlers into their car seats and drowned them by submerging the car in a South Carolina lake. I wondered in what kind of a home she grew up and if she was mentally ill. Newspaper accounts provided little information but her trial provided the answers. She came from a very physically and sexually abusive home and she was indeed mentally ill. In her case, a severe episode of depression was the precipitant.[1]

Some internal pressure based in mental illness may have pushed Kinkel, Weston, and Smith "over the edge." But many who accept that concept in those cases are wary to accept the idea that other crimes are also the result of mental illness, neurological illness, and an abusive upbringing. Crimes like serial murder and murder-for-profit mostly seem to be inspired by evil, not illness.

Ted Bundy was a handsome, intelligent man who had attended law school. With a combination of good looks and charm, he enticed dozens of unsuspecting women to accompany him on dates that led to their deaths. Jeffery Dahmer collected young men, plied them with drugs and drink, and then killed them and ate parts of their bodies. On a murderous spree over a period of weeks that extended from the Midwest to Florida, Andrew Cunanan ultimately killed Gianni Versace and then committed suicide. I examined Bundy, have spoken to Dahmer's mother, and predicted during a TV interview Cunanan's suicide. In each case, as far as I was able to determine, issues existed that were untouched by the journalistic treatment of these nationally notorious killers.

There are also regionally infamous serial killers: Joel Rivkin in New York, who killed sixteen prostitutes during a four-year period and dismembered several of them in the basement of his mother's house. Douglas Clark, known as the Sunset Boulevard Killer, also preyed on female prostitutes. He decapitated one and maintained the toothless head in his refrigerator. William Bonin, the California Freeway Killer, picked up male hitchhikers, raped, hog-tied, tortured, and garroted them. Normal parents? Normal brains? I think not. Abusive experiences, mental illness, and neurological deficits interplayed to produce the tragedies reported in the newspapers. This complex of factors is the subject of this book.

Unifying causal themes link planned murder with fury-inspired, random murders. Murders committed in pursuit of money or drugs are not different. The motivation of a "rational" killer seems clear, such as that of a bank robber in Richmond, Virginia, who shot two people or a convenience store robber who killed a clerk in Modesto, California. Yet, the same complex of factors underlies the act of homicide. Attacks on strangers are substantially the same, etiologically, as attacks on friends. For example, apparent hate crimes like Russell Henderson and Aaron McKinney's killing of Matthew Shepard, a young homosexual in Wyoming, I found originate in a background of abuse, mental illness, and brain damage.

During the past twenty-five years, I have examined about 150 murderers, including most of the killers named above. Regardless of the classification of the killing, I believe most killers kill for the same reasons.

Base Instincts

CHAPTER 1

The Theory of Violence as Taught by Louis Culpepper

On a pleasant Sunday morning, two correctional officers led Louis Culpepper to meet me in the medical office of the Augusta, Georgia, jail.[1] The jail had been Culpepper's home for fourteen months while he awaited trial for sexually molesting his six-year-old stepdaughter.

He wore sneakers and an orange jumpsuit with short sleeves that revealed his biceps, defined sharply by hours of exercise with prison weights. He was a ruggedly handsome thirty-year-old man with a neatly clipped moustache. He demonstrated social grace in greeting me. When introduced by his lawyer, he looked me in the eye, smiled, and grasped my extended hand and shook it, warmly. In appearance, he resembled the actor Burt Reynolds. The outer man was fine. What horror could have so deformed the inner man?

I had been engaged by his lawyer to answer that question, to uncover facts that might explain and mitigate the loathsomeness of what he had done. The purposes of the defense team and mine differed but the paths to them coincided. The defense wanted him to receive the shortest possible sentence and sought mitigating factors. I wanted to understand the pathogenesis of violent crime. I was an expert for the defense. As a neurologist, a medical specialist in diseases of the brain, I evaluate not just motor and sensory functions but cognition and behavior as they are major outputs of the brain.

A few weeks before he began to molest his five-year-old stepdaughter, Culpepper had been badly injured in a near-fatal motor vehicle accident. He was driving home, after work, when an intoxicated driver crossed the median divider and struck Culpepper's car, head-on. Not wearing a seat belt, Culpepper was catapulted through the windshield, head first. Unconscious, he was rushed to the hospital. A CAT scan of his brain showed hemorrhages with large clots in his right and left frontal lobes, the portion of the brain that lies behind the forehead and in front of the ears. There was severe swelling of the brain adjacent to the clots—a very dangerous situation. Because the brain is soft and the skull around it is hard, the pressure of the clots and swelling in his frontal lobes was transmitted backward and downward, like toothpaste in a squeezed tube, toward the part of the brain that is behind the nose and the mouth. That is where the vital centers that control breathing and heart rate are located. His life was hanging by a thread. A neurosurgeon operated on him that night, removed the clots, relieved the pressure, and saved his life.

Culpepper made a gradual but steady recovery over the course of a few weeks and was discharged from the hospital having no motor or sensory deficit, except that he had lost the sense of smell. His ability to read, write, speak, and remember were unimpaired, though he had no recollection of the accident or of the two weeks following it, when he had been in the intensive care unit.

Culpepper told me that during the period of recuperation at home, before his doctors had permitted him to return to his job, he had been left alone with his young stepdaughter while his wife had gone out to do the grocery shopping. He dozed off on the living room couch and he developed an erection in his sleep. His stepdaughter was standing next to the couch when he awoke. He thought that she showed sexual interest in him by staring at the bulge in his trousers, and he became excited. He unzipped his pants and placed her hand on his genitalia and directed her to stimulate him to orgasm.

This was the first of many sessions with the little girl that extended over the course of the following year. He taught her to stimulate him orally and he touched her genitalia with his fingers but did not penetrate her. After he climaxed, he would kneel on the floor with the child. He led her in prayer, begging the Lord for forgiveness.

Ultimately, his wife came home unexpectedly and witnessed her child stimulating him. His wife was distraught and told her minister, seeking his guidance. He called the police. Culpepper was arrested and charged with criminal sexual conduct with a minor and endangerment of a child. He faced ten to twenty years in prison for this, the first arrest in his life. While he was in jail awaiting trial, his wife divorced him.

When I examined him, he was calm, collected, and apparently rational, except that he thought his six-year-old stepdaughter had made sexual advances to him. He was certainly not psychotic or depressed. He spoke about his likely incarceration in prison without anxiety. He said that his appetite for food and his ability to sleep while he was in jail were normal, unaffected by his circumstances. He knew that child molesters are targets for other prisoners who beat and kill them. Yet, he was not agitated or trembling at the prospect of a long term in prison.

He discussed his behavior and urges without expressing shame and fear. He said, "What I did was wrong but modern society makes too much of sex with a child. I did not abuse her. I did not force her to have sex with me, and I did not hurt her. If my wife had not told the minister, I could have straightened things out. I could have gone to therapy, and I would still be married."

He suggested that his young stepdaughter had initiated the whole thing by looking at him in a "sexy" manner, thereby indicating her desire to be sexually involved with him. When I told him that it was impossible for a six-year-old to initiate sex with an adult, he argued the point. He had no concept that he might have harmed the child. He did not see himself as others saw him and did not seem to care about the consequences for himself or the child. He did not express verbally, with facial expression, or body language the emotions that should have accompanied what he said. The flatness of his affect reflected his lack of emotional feeling.

Culpepper's lawyer had told me that when Culpepper was a

child he had been sexually abused by his father and grandfather and some of their friends. He had been beaten, penetrated, and forced to perform fellatio. His abuse had ended in his early adolescence, long before he began to abuse his stepdaughter but extended through most of the formative period of his brain's development. He had finished high school, worked at the same job for eight years, and married a woman who had a daughter by a previous marriage. His wife told investigators that their marriage had been stable, warm, and nonviolent. He had been reliable and caring to her and to her young daughter. They seemed to be an ordinary family.

His lawyer thought that the history of Culpepper's abuse in his childhood was not relevant to the case. After all, he pointed out, it had ended without Culpepper's having committed any offenses for the next fifteen years. He thought that Culpepper's brain injury was the cause of his crime because the brain injury had occurred shortly before Culpepper first molested the young girl.

I suspected that causation was not so clear-cut. I discovered that the injury did not initiate Culpepper's pedophilic urges. Culpepper confided to me that he had fantasized about having sex with a child for years but, before the accident, he had never acted on his impulses.

Culpepper was one of the first criminals I had seen whose damage occurred in adulthood, had a clear date of onset, a cause, and a well-defined location in the brain. My examination indicated that he had sustained two kinds of damage: developmental and physical. His mind had been damaged by a horrible childhood, and his frontal lobes had been ruined in an automobile accident.

Culpepper crystallized for me two of the disparate elements I had regularly found in violent individuals. The sexual abuse he suffered in his childhood was necessary to establish his pedophilic urges but had not been sufficient to set them off. Having been a victim of childhood sexual abuse imprinted on him the urge to sexually victimize children. He had carried this imprint and the impulses it generated throughout his life. Until the car accident, those impulses had been inhibited by the frontal lobe of his brain.

Absent the prior abuse, Culpepper, like the vast majority of people who have sustained brain damage, might have been emotionally blunted, impaired in his concerns about the consequences of his acts, limited in the capacity to be self-critical, and may have even had reduced sexual inhibitions. He would probably not, however, had the specific desire to sexually abuse children. By interacting with each other, his sexual abuse in childhood and this traumatic brain damage had transformed him into the perpetrator of a sexual crime against a child.

This case and cases like it led me and my colleague, Dr. Dorothy Lewis, to develop a new theory, one that has the capacity to explain violent crime. According to this theory, the elements we had found in other murderers interact in the same manner as they had in the case of Louis Culpepper. It is the interaction of childhood abuse with neurologic disturbances and psychiatric illnesses that explains murder. Abuse generates the violent urge. Neurologic and psychiatric diseases of the brain damage the capacity to check that urge.

CHAPTER 2

Murder on the School Bus

The insight into the etiology of violence that Louis Culpepper provided stimulated me. I began to rethink several cases that had puzzled me before.[1] An obvious candidate for reconsideration was the case of Cynthia Williams, the first murderer I had ever met. She certainly did not fit the ordinary profile of a murderer. She was a girl in her early teens.

Cynthia had told me she had not felt well on the day of the incident and had not wanted to leave home that day to attend her seventh grade classes, but her mother insisted that she take the school bus from her intercity apartment to the Fairmont Middle School in the upscale Northwood section of town. Cynthia had argued with Mona, a big, tough, loud, eighth-grade girl the day before. Mona had said, "I'm going to get you." She knew Mona and her brothers because they lived in the same neighborhood and were constantly in trouble with the police. At the thought of actually fighting Mona, Cynthia said she felt sick to her stomach. Mona could beat her senseless without trying. What could she do?

Cynthia wanted desperately to stay home but feared the beatings she would receive from her mother and her mother's boyfriend if she skipped school that day. She remembered her mother's demand that she fight back. "If anyone starts up with you," her mother said, "beat them up. Even if you lose, if you put up a good fight, you will get 'respect.' People will know they cannot just mess around with you and expect not to get hurt. Don't come to me complaining that someone hit you, or I will beat you." So, Cynthia took a knife from the kitchen that day and hid it in the sleeve of her winter coat, just for protection.

She seldom wanted to go to school. She had been a "slow learner." She did not understand many of her classes, could not always follow what the teachers said, and her grades were poor. The class she hated the most was gym. She always felt uncomfortable during gym, especially in the locker room. She was afraid the other girls would mock her for the bruises on her back and thighs caused by beatings delivered by her alcoholic mother and her mother's boyfriend.

She sometimes thought that she heard someone call her a bad name. Afraid to confront anyone because this could lead to a real fight, Cynthia would tuck her head down and pretend she had not heard anything, but it made her angry that they laughed at her. On one occasion she was so angry that she wheeled around to confront her tormentor, but no one was there. She was both relieved and frightened. There was no need to fight, but she had heard a sound that was like a voice, laughing. The person who had laughed at her may have been hiding. She began to carry a knife to school to protect herself, especially when she sensed that she would need extra protection. Mona's threat made her feel that she needed protection that day.

Mona was already on the bus and was talking loudly to other girls in the back when Cynthia got on. Cynthia took a seat near the front and tried to slouch down so that Mona could not see or notice her. It worked. When they reached school, Cynthia hurried away and went to her homeroom before Mona could get off the bus.

Cynthia later said that she felt a sense of relief when she arrived at school safely, yet she immediately began to fear the ride home. What if she could not avoid Mona going home on the bus? She could think of nothing else all day and sat silently in her classes, totally consumed by thoughts of Mona, fighting, escaping.

Toward the afternoon, Cynthia started to feel ill. She had a pounding headache, and she felt nauseated. She had a sense that something bad was going to happen. She was terrified of Mona. She wanted the principal's permission to leave school early and go home alone because she felt sick. She asked to be allowed to walk home because she did not know how to transfer on the public buses to get from the school to her home. The principal insisted that she take the school bus rather than walk. He said it would not be safe for her to walk so far, especially if she felt ill.

When the bus arrived, she got on and took a seat at the very back. She felt as if everyone were looking at her. The other children said later that Mona entered the bus after Cynthia sat down. Mona saw her and began the long walk from the front to where Cynthia was sitting.

Cynthia said that she was dizzy. She looked down the aisle of the bus at Mona. The other children's faces changed in size. She sensed the odor of a dead rat. "I'm going to get you," Mona said. Cynthia had no further recollection.

The other children reported that Cynthia stood up to face Mona. With a single movement that looked like a punch, Cynthia's knife, previously concealed in her right hand, stabbed into the left side of Mona's chest. The victim, whose heart had been pierced, looked amazed. She stood still for several seconds before twisting to the side and collapsing backward into a seat. Bloody bubbles appeared in her mouth and a rapidly enlarging bloodstain appeared on her chest. The dying victim gasped several times. Her eyes looked empty. She stopped moving. One child cried out, "She's been stabbed! She's bleeding!" There were screams. Some approached the unresponsive girl to see what had happened and others stood on their seats, craning their necks to observe. The bus driver entered the bus and pushed toward the back to assess the situation.

The children seated in the front of the bus told the police that immediately after she had stabbed Mona, Cynthia looked as if she heard and felt nothing. She acted as if nothing had happened. Her face was blank, her eyes without expression. She did not run but walked, slowly, to the front of the bus, got off, and walked away.

Cynthia was not able to remember the stabbing or to account for the two hours that followed. At dusk, she called her mother from a phone booth a mile away from her house. She was picked up by her mother and driven home. She was sleeping in her bed when the police came later that evening and took her to the juvenile detention center.

This murder, publicized by the local newspaper and television station, sent shock waves through Northwood, a neighborhood generally immune from violence. The entire community was stunned. People placed more locks on doors and windows, purchased guns and attack dogs, and demanded metal detectors and security guards at the school. The media were decried for depicting too much violence.

How could a young girl kill another young girl on a school bus without a motive? There was much detailed reporting about the murder—eyewitness accounts and an interview with the principal. No attention was given to the question, "What made thirteen-year-old Cynthia kill?"

Only 6 percent of the population causes the vast majority of violent crime—about 70 percent.[2] Any behavior that is so disproportionally restricted to such a small segment of the population is abnormal, by definition. If someone's behavior is abnormal, could that person have something wrong with his or her mind, that is to say, the brain? There was very little disposition among the citizenry to understand why this incident had happened, yet there was some awareness of abnormality implied in the statement of a Northwood housewife, "There's got to be something wrong with her. Can you imagine a child doing that?"

In the medical school dining room, the discussion was not particularly sophisticated. My colleagues around the lunch table seemed satisfied by diagnosing Cynthia as a sociopath or as having an antisocial personality disorder. True, Cynthia had acted antisocially. Yet I thought that antisocial personality disorder was not the explanation for her antisocial act but the phenomenon that needed explanation.

The murder was terrifying. What response could there be to such a dreadful act? The prosecutor said that he would try Cynthia as an adult. The court asked for a psychiatric evaluation.

Cynthia was seen at the juvenile detention center by Dr. Dorothy O. Lewis, a child psychiatrist who was then working at the juvenile court. After examining her, Dr. Lewis suspected she had neurological problems, brain damage, and perhaps seizures. Dr. Lewis asked me to see the girl. As a neurologist on the local medical school faculty, I took part in a thorough investigation that involved interviews with Cynthia's family members and school authorities, as well as a review of her medical records.

Cynthia's lawyer arranged for me to examine her at the juvenile detention center. There, in a small room with a large desk, I expected to see some kind of monster whose appearance matched the enormity of what she had done. Escorted to the door by a guard, she entered the room alone and sat down on a stiff, armless chair, her hands in her lap, her head bowed.

She did not look like a murderer. It seemed impossible for her to have done any harm to anyone. Cynthia was a prepubescent, African-American girl, about five feet tall and weighing no more than eighty pounds. Her posture was self-effacing. Head down, she did not or would not look at me, averting her gaze. She did not establish eye contact more than two or three times during the entire period of the examination. In response to my questions, her little-girl voice emitted barely audible, single words, mostly monosyllables. At first, she gave me very little information about herself or the stabbing. The descriptions provided by other children on the bus were far more complete. In subsequent interviews, when she became more comfortable with the doctors engaged by her lawyer, she provided a more complete picture of her subjective state on the day of the incident.

The most reliable medical information came from Cynthia's hospital records. Cynthia's charts showed that her abuse began in utero. Her mother put Cynthia's future health at risk in several ways: her mother drank heavily, had syphilis, and had abnormally low thyroid function (hypothyroidism). Any one of these maternal factors can damage a developing fetal brain. In addition, Cynthia's entry into the world was difficult. The delivery was traumatic enough to severely deform her skull and break her collarbone. These injuries were noted in the hospital's records at the time of her birth. Such injuries are associated with an increased risk of brain damage.

Cynthia's early childhood was full of accidents, injuries, and illnesses. The hospital records documented more than thirty visits to the emergency room, mainly for trauma. On at least one occasion, Cynthia was seen for a concussion severe enough to require the treating physician to order skull Xrays. This indicated that the physician thought it was possible that the blow had fractured her skull. Such a high number of emergency room visits is characteristic of abused children whose parents may bring them for medical attention but provide misleading historical information to the doctors, never mentioning abuse. "She fell down the stairs, fell out of a window, tripped on a rug and hit her head, was hit by a car, fell out of bed. She is always hurting herself." Those are typical statements of abusive parents.[3]

The question in this case was whether any of these potential injuries to the girl's brain were in any way relevant to the stabbing. For example, I wondered if the stabbing were the direct result of an epileptic seizure. A complex partial seizure[4] could have explained why she could not remember the event and its aftermath and why she was sleepy. The peculiar smell and visual distortions that she reported having just before the incident are typical of the symptoms that precede complex partial seizures. The stab could have been an automatic motor act caused by the firing of epileptic nerve cells. However, the duration of the buildup, the event, and the aftermath were too long. Directed violence is so rarely caused by a seizure that even a poorly planned, but not entirely spontaneous, stabbing like Mona's murder by Cynthia would almost never be the result of an epileptic seizure. Yet she may have had complex partial seizures before the incident.

Her mother described brief periods during which Cynthia stared blankly and acted oddly and for which she had no recollection. These lapses could have been seizures.

It was also possible that she had a kind of migraine syndrome with periods of confusion and irritability. A child's migraine is often characterized by these manifestations. This would account for the headache and the nausea Cynthia described before the incident, but it is probably fair to say that murder is never the direct result of a migraine headache. It would be much less likely that migraines directly caused the murder than epilepsy. Still, some migraine sufferers become irritable before, during, and after their headaches. Conceivably, irritability could have contributed in some way to the act.[5]

Alternatively, maybe she just wanted to kill her tormentor, lest Mona get her first. Could she merely have been lying about forgetting the details to excuse or to mitigate her actions? Even if this were true, it would not explain why Cynthia killed Mona at that time. Without a medical-scientific approach, the only explanation becomes a moral-religious one: the devil made her do it.

Even in those early days, before I had fully developed my examination techniques for people with behavioral disorders, I found several abnormalities on my neurological examination of Cynthia. She was microcephalic; that is, the circumference of her head was significantly below average. Because the head size is determined by the brain size, microcephaly always indicates that the brain is small. Of people with small heads, 95 percent have intellectual deficits. Microcephaly can result from many factors present during pregnancy, including infections or maternal exposure to drugs and alcohol. Her thin upper lip and shallow philtrum (the vertical indentation between the nose and upper lip) were signs of fetal alcohol effect.

She had marked choreiform movements (jerky, irregular, involuntary movements) visible when she stretched out her arms and spread her fingers. Her motor coordination was abnormal for her age. She could not carry out certain motor tasks that normal seven-year-olds can. For example, she could not skip or walk a straight line. These were indications of neurologic abnormality, but did they explain why she committed murder?

It is overly simplistic to assert that she was violent because she was neurologically impaired. Not all people with brain damage behave violently. In fact, only a small minority of people with brain damage ever become violent. This was shown to be true of Vietnam veterans with penetrating wounds of the brain; only a small number acted violently in the years subsequent to their injury.[6] There had to be something different about this brain-damaged child who had committed murder from the majority of brain-damaged children who are not violent. What could this crucial difference have been?

Since Cynthia was the first murderer I had ever examined, I did not know then that there are two other factors that are associated with violence besides brain damage. The most vicious criminals have also been, overwhelmingly, people who have been grotesquely abused as children and who have paranoid patterns of thinking.

At that time, I, like most people, believed that violent criminals and delinquents were just like the rest of us except that they were "bad." They had been inadequately supervised by their parents and not disciplined enough. Their homes were troubled, their parents were irresponsible, and they had developed inappropriate patterns of behavior. It has been amazing to discover that the quality and the amount of "discipline" these individuals have experienced are more like that of a prisoner in a concentration camp than a child at home.

Abuse has been the most surprising and possibly the most significant finding of the research into the causes of violence. Child abuse has devastating psychological consequences for children and for the adults that they become. This finding has now become conventional wisdom. What is new is the growing perception that prolonged child abuse permanently changes the anatomy and function of the brain.[7] Child abuse has thus moved away from the purely sociological and psychological realm of interest and into the neurological sphere. Abuse can potentially damage the brain by direct trauma. More insidiously and pervasively, abuse alters the basic developmental anatomy, physiology, and functioning of the brain.

I did not limit my examination to the neurological, but performed a general medical examination as well. The skin and the scalp are repositories of information that often confirm the history and sometimes open new lines of questioning. Cynthia had been abused. She said that she had been beaten by her mother and her mother's boyfriend, and she had the scars and bruises on her back and thighs to prove it. In all likelihood, I now realize in retrospect, the lapses in her memory for the stabbing and its aftermath and all the other lapses in her memory that her mother had described represented periods of dissociation. I did not even consider dissociation when I saw her twenty-five years ago. I did not know how to relate her experience of abuse to the murder.

Most people who have been abused are not violent. As a matter of fact, about 90 percent of people who have been so badly abused that public agencies have been called in to intervene do not require the efforts of public agencies to protect their own children and do not become violent criminals.[8] The capacity of abused people to lead relatively normal lives is a testament to the resilience of the human spirit, i.e., the plasticity of the human brain. Yet a large number of formerly abused people do become violent and dangerous to society, and there is an undeniable direct link between the experience of abuse and later violence.[9]

Obtaining a history from Cynthia had been like trying to run through a field of molasses. The difficulty seemed to be that she did not trust me even though I was on "her side." Her pervasive sense of being endangered was out of proportion to any real threats in her environment. It was her response to such thoughts that had caused her to go around armed and ready to attack the unarmed girl who had merely verbally threatened her. The stabbing was an excessive response to a feeling of being imperiled.

I wondered if it were situationally appropriate for a poor child from a bad neighborhood to carry a knife to school for protection. On what basis could I decide this? Such behavior would be patently abnormal in a middle-class child from a good neighborhood.[10] Cynthia's thinking was paranoid, perhaps not fully delusional but certainly mistaken and had directly contributed to the murder. Still, most people with paranoid thoughts and even fully developed paranoid delusions are not usually violent.[11] Why were Cynthia's paranoid thoughts so lethal? Paranoid thoughts are commonly manifested by people who are psychotic, as in schizophrenia, severe mania, and severe depression. Could Cynthia have been psychotic? There was no evidence of systematized delusions or hallucinations on my examination. Her reluctant, terse responses were appropriate and rational. Yet she was very paranoid.

Cynthia's paranoid thoughts were not occurring in a brain that was otherwise normal. Her brain had been damaged by alcohol in utero and by abuse throughout her childhood. Now I know that the three elements that were present in this little thirteen-year-old girl—neurologic damage, abuse, and paranoid thinking—are almost always present in adult murderers on death row. Each of these factors alone may not have had sufficient force to impel little Cynthia to violence, but together they created a devil's brew that made her into a murderer.

When Dr. Lewis and I studied fourteen death row inmates who had been condemned for committing murder before they were eighteen, we found the same mixture of factors. There have been isolated reports of murders committed by even younger children, like Robert Thompson and John Venables in England, ten-year-olds who lured a toddler to his death. Both boys came from dysfunctional families where violence was prevalent.[12]

Twenty years ago, there was no evidence that the triad—neurological deficits, paranoid thoughts, and abuse—were prevalent among violent people. In fact, common wisdom then held that this triad was *not* prevalent.[13] Dr. Lewis was the maverick who believed that this triad existed in violent people and convinced me to join her in testing this hypothesis.

There was only one way to find out whether neurological deficits, paranoid thoughts, and abuse were common features of violent criminals, and that was to examine other violent people.

It was not so easy to find violent people, except in a place of confinement, and neither defense lawyers nor the correctional authorities were especially enthusiastic about permitting me to study prisoners. I therefore gratefully accepted Dr. Lewis's invitation, a few years later, to participate in her study of violent and nonviolent delinquents. Dr. Lewis and I evaluated ninety-seven boys and twenty-two girls who averaged fifteen years of age in the only "reform" school in the state that held children under the age of sixteen, the Long Lane School. We found that violent behavior correlated with neurologic abnormality, paranoia, and abuse. We reported this in a series of peer-reviewed papers for professional journals.

At about the time we were winding up our work at the reform school, Dr. Lewis was offered and accepted a full-time position as professor of psychiatry at New York University. Soon after arriving, she made contacts with the chiefs of the forensic services at Bellevue Hospital. Bellevue has medical and psychiatric inpatient facilities where prisoners are cared for and evaluated. She arranged for us to see young men incarcerated on these wards. The plan was to check my findings from the reform school delinquents where I had known who was violent before I performed an examination. I was anxious to perform neurological examinations before knowing the history, to determine if violent offenders still appeared to be neurologically impaired as compared with the less violent offenders.

I was in a very happy frame of mind at the time of my first visit. I had not visited Bellevue, that ancient landmark, since my medical school days. Then, the Columbia College of Physicians and Surgeons, my alma mater, had operated several services there known as the old First Division. My happiest memories of medical school are not the gleaming halls of the Presbyterian Hospital in majestic Washington Heights overlooking the Hudson River, but the cockroach-infested, vibrant wards of Bellevue, filled with alcoholics from the Bowery and the impoverished from the Lower East Side and the areas between. No magnificent apartment-condominium cooperatives lined New York's First Avenue then, only the bleakest tenements and cold-water flats.

The hospital's facilities and personnel in those days never could handle the crush of patients whose beds spilled out into the halls. The only thing that stood between these patients and death was the valiant, unremitting outpouring of energy, devotion, and time spent by the interns, residents, and medical students working with them. There was a real feeling of romance at Bellevue. Nothing seemed to work. One's ingenuity, charm, wit, and daring were necessary to obtain Xrays and electrocardiograms (EKGs), to win the cooperation of recalcitrant technicians and overburdened nurses on our patients' behalf.

We medical students did the blood counts, the urinalyses, the analyses of body fluids ourselves. We drew blood, set up IVs, and often transported patients for tests. As a fourth-year medical student, I successfully treated a man on the surgical ward who had diabetic acidosis. I administered fluids and insulin, guided only by glucosticks and Acetest tablets for measuring urine sugar and acetone. The resilient and lucky patient survived the night and I experienced one of the most deeply satisfying moments of my career.

I was brimming with memories of my halcyon days when Dr. Lewis and I entered the vestibule of the old Bellevue Psychiatric Service on Thirty-first Street. High vaulted ceilings lined with tiles magnified each sound. Marbled columns and worn stairs with decorative handrails gave evidence of an elegance that never existed there but was proposed by some antediluvian architect.

The halls were filled with shrieks, screams, and demented laughter. As we waited endlessly for the elevator, the cacophony increased in volume and the number of voices contributing to it gradually augmented. It was disturbing; I had not recalled this degree of bedlam. Dr. Lewis, as if reading my mind, said calmly, "You know, those sounds are not from patients. These are not wards. These are administrative offices. They are having a coffee break."

The inmates of the forensic psychiatry service at Bellevue were all prisoners who were suspected by a judge of being mentally ill. The ones found to be ill were ultimately transferred to a state mental hospital; the others would be incarcerated in the jail on Riker's Island. Most of the patients I examined were derelicts, tramps, and the homeless of New York.

At Bellevue, I had the only close call regarding my personal safety in the twenty-five years I have worked with violent criminals. As I mentioned, my neurological examination was to be done first, before I obtained the history or reviewed the chart. I purposely did not know who was violent. A distracted secretary selected patient-prisoners for me to examine on the basis of age alone. She assigned various empty offices to me in which I could examine the patients.

On this occasion, the office she had chosen was located at the end of a secondary corridor, a cul-de-sac off the main hallway. There were several unoccupied rooms between "my room" and the main corridor. The prisoner-patient, in handcuffs, whose status was unknown to me was accompanied by a guard who led him to "my room." The handcuffs were removed for my examination and the guard who brought the prisoner returned alone, through the deserted corridor to the main corridor that was bustling with activity.

The subject was a distracted black man with huge muscles, about six feet, eight inches tall. As the guard left, the patient-prisoner inappropriately chose to sit in a chair next to the office door. I asked him to sit on the examining table. It was located on the wall opposite from the door and would have allowed me to be closer to the door than he. He did not move nor did he speak. His eyes darted around the room, his body jerked; he turned his head as if checking to see if anyone was coming, and he did not respond to my questions. He made me feel uncomfortable. I felt that I could not just initiate an examination, so I tried to start a conversation by asking what I thought were very nonthreatening questions.

"How old are you? How many brothers and sisters do you have?"

He would not answer these questions but smiled strangely and inappropriately, his eyes still darting about the room, his body twitching as if energized by some unseen loose electrical connection.

He began to ask me questions. The questions were benign but the tone was menacing. "What you want to talk to me fo'? What you want to know?" I tried to explain what a neurologist is and why I wished to examine him. He acted as if I were not speaking to him.

His constant inappropriate smile was as disconcerting to me as his physical size was intimidating. I began to feel very nervous. He was between me and the door in an office at the end of an empty hallway. I had no way out. There was a telephone on the desk that I could have used to call for help, but I did not know the number to dial. I considered calling the operator at Bellevue, but in my mind's eye, I saw her having a noisy coffee break near the entry of the hospital. By the time I explained to her where I was and what I wanted, I might have been weltering in a puddle of my own blood with my head inside out. This thought was sobering. I was frightened. I could have shouted but might not have been heard unless I screamed in a very loud voice. I was embarrassed to cry out for help in this manner.

I was about to pick up the phone and give it a try when, providentially, guards suddenly came in, handcuffed the subject, and led him back to the cells. This ended the interview and I breathed a tremendous sigh of relief. I tried to determine the source of my salvation. It turned out to be the disorganized secretary. It seems she had gone to an adjacent room to retrieve a form. She recounted the situation to me afterward in an austere monotone: "I didn't like the vibes, so I got help." I told her I had been on the verge of calling for help by telephone. "Oh," she smiled, "that phone doesn't work. The line's been dead for months."

I have never again been casual about my personal safety in prison. Ironically, dealing with the most dangerous prisoners on death row has probably been safer as security is usually very tight. Prisoners are often shackled and guards are never far away or out of sight. As a student, I had always looked upon myself as a rescuer at Bellevue, yet I had required rescue there.

Ultimately, I examined forty young men who were prisoner-patients at Bellevue Hospital. The association of violence with neurologic abnormality was confirmed. Then Dr. Lewis and I examined thirty-two nondelinquent children at an after-school recreational program in New York (the Police Athletic League) and compared them to an age, race, sex, and socioeconomically matched group of delinquents. In all these studies, there was a strong tie between violent behavior and neurological abnormalities, paranoid thoughts, and the experience of severe, prolonged, physical abuse.

I did not know how abuse and brain damage led to violence twenty-five years ago. Consequently, Cynthia's murder of a schoolmate seemed terrifyingly unpredictable and random. From my present perspective, after twenty-five years' experience examining murderers, it now seems clear that brain damage crippled Cynthia's judgment. Violent urges engendered by abuse and paranoid thoughts were strong. Her social inhibitions were weak. The weakness of her capacity to inhibit her violent inclinations led to murder.

CHAPTER 3

Murder by Abuse

The findings of the studies Dr. Lewis and I conducted on violent delinquents were of interest to many defense attorneys, some of whom asked us to examine a group of death row inmates whose execution dates were imminent. The lawyers who requested my services hoped that I would find abnormalities that could mitigate the death sentences of their clients to life in prison without parole.

I really wanted to examine the most violent individuals to determine whether they shared the neurologic and psychiatric vulnerabilities and the experience of abuse in childhood that Dr. Lewis and I had discovered in the delinquents and other violent prisoners we had evaluated up until that time.

Colleagues on the tweedy university campuses on which I have spent the bulk of my professional career (I was a professor of neurology at Yale and chair of neurology at Georgetown) have often asked me how I can visit so many prisons, meeting, shaking hands with, and spending several hours alone with some of the most violent criminals in the country. I answer that if one wishes to examine violent people, one must be prepared to make house calls. Violent people in custody seldom come to the plush offices and examining rooms of the Georgetown University Hospital. Yet not all my "carriage trade" patients have been completely nonviolent. How violent must one be, how often, and over what period of time to be considered violent? To avoid the ambiguities of classification, I decided it would be best to examine individuals society had unequivocally labeled "extremely violent." That is to say, death row inmates. My first foray to death row provided a number of unforgettable experiences and vivid impressions that advanced my understanding of what makes killers kill. This is the story of one murderer, Bobby Moore, whom I encountered on my first trip to death row.

The story of his crime, as reported in the local papers, began in a parking lot. The parking lot was sparsely dotted with cars and no other people were outside in the hot, southern, summer sun when Laura Foss returned to her car from a rural grocery store. Her arms were laden with packages and she was having a difficult time manipulating them because she was seven months' pregnant. She was placing the groceries she had just purchased in the back seat of her blue Toyota when she was approached by two men, one pointing a gun at her. One of the assailants told her to sit in the driver's seat. The man holding the gun sat next to her on the passenger's side, directing her as the second man followed them in his own car.

A few hours later, Laura's partly naked body was discovered in a wooded area, lying face down, a bullet hole in the back of her head. Someone had fired a pistol at close range. A torn check drawn on Laura's bank account was found about nine feet from her body. It was made out to "John Doe" in the amount of $20,000. It was dated on the day of her murder and bore her signature.

An autopsy on Laura revealed an abrasion below the right side of her chin, probably caused by a punch. There was a bruise on the right nipple, likely caused by oral suction. Scrapes and large bruises on her shoulders and neck indicated beating and choking in a mortal struggle. The fresh semen inside her vagina indicated that she had been raped. Whoever shot Laura was later determined to have killed a police officer in a nearby town, using the same pistol.

Later that evening, Sally Parsons, a convenience store manager, first became suspicious of two men and then became alarmed by them. She wondered why two males driving a blue Toyota repeatedly slowed down as they passed her store without stopping. After driving by three or four times, the two men parked and entered the store. They lingered in the aisles for a long time and seemed a bit on edge. The suspicious men, who pretended to be carefully considering the purchase of ordinary items like candy, drinks, magazines, and toiletries, glanced frequently at two other men in the store who were chatting over a cup of coffee. Finally, they settled on two unusual items—a small teddy bear and a pair of socks. They paid for these and left, much to Parsons's relief.

Yet Parsons felt uneasy as she noticed that they continued to drive by the store, slowing down as they peered in. Parsons believed that the men driving by were waiting for the store to be empty and that she was in danger. She was alarmed enough to ask the last customer, who was still lingering over his coffee, to call the police on the pay phone in the store while she kept her eye on the two men in the blue Toyota.

Soon after the customer called the police, the two suspiciously acting men again parked in front of the convenience store and came in. They remained in the store, once again looking over various items but darting their eyes around the store evaluating the other patron.

Within ten minutes after the pair had returned to the store, a police officer, responding to the call, arrived at the convenience store in his patrol car and parked next to the only other car in the lot. By radio, he requested a check on its license plate. The car was registered in the name of Laura Foss. Her murder had not yet been reported. Laura Foss was a white female. The two suspicious men were black. The officer grabbed his shotgun and quickly entered the store. He exchanged a glance with Sally Parsons who affirmed with a nod the location of the two suspicious men. The officer approached them and politely but firmly requested that they step outside with him to answer a few questions. Without a word, the two slowly preceded the officer to the door.

One exited and walked directly toward the blue Toyota, the other followed, and the officer was the last to leave. The two men stood between the cars. The police officer's car blocked Parsons's view. A loud gunshot rang out. The two men jolted forward in their car and with squealing tires sped away.

Sally Parsons and the customer who had called the police for her ran out of the store, and on the far side of the patrol car they saw the officer lying face down on the ground in a growing pool of blood. A pistol was next to his motionless body, but his shotgun was missing and his holster was empty. Panicked, they ran back into the store, called the police and told them what had just happened. A general alarm was broadcast across the county by police radio.

About half an hour later, in a nearby town, police officers spotted a blue Toyota with two black men driving erratically at high speed. Police officers gave chase, and the men fired several rounds at the police. The two men abandoned their car in an orchard and attempted to escape by foot. The police officers followed and caught the men in a pasture next to the orchard. In the blue Toyota, police recovered the guns belonging to the dead officer, Laura Foss's groceries and handbag, and the teddy bear and socks purchased at the convenience store.

Bobby Moore was one of the two killers. In his tape-recorded confession, he revealed that on the day of the killings, he had been drinking brandy at a bar with an acquaintance, Thurmon Johnson, the other man implicated in the murders. They consumed enough brandy to get themselves drunk. Their minds cloudy, they left the bar with the intention of holding up a convenience store.

Bobby was carrying his mother's pistol. He claimed that he often carried the pistol for his own protection without having any particular need for it. Contrary to the police department's theory that the crime was long premeditated, Bobby explained that he just happened to have the gun that day and that he had not planned to rob a store when he left home that morning. If he did carry his mother's gun for the purpose of robbing a store, it would have been the only clear act of planning ahead on Bobby's part in the events of that day.

According to Bobby's confession, Thurmon Johnson did not want to use his own car for the robbery and proposed that they steal a vehicle instead. Bobby, who did not own a car, went along with this plan. They drove from the bar where they had been drinking to a nearby grocery store. They waited in the nearly empty parking lot. They saw a pregnant woman leave the store. As she struggled to open her car door, they approached her and forced her at gunpoint to drive with Bobby.

As the two cars drove off, the terrified woman begged not to be harmed. "Do whatever you like but don't hurt me!" she said. Bobby promised not to hurt her, but when they arrived at a lonely orchard, Johnson said he wanted to kill her because he was afraid that she could identify them. Bobby said that he just wanted to tie her up and use her car, not kill her. They had not considered what they would do with the car's owner before they stole the car.

In their confessions, both men denied raping Laura Foss, and nothing could be definitively proven in court regarding rape since a DNA test of the sperm was not performed. Consequently, neither Bobby nor Thurmon Johnson was charged with rape. Yet the police, the prosecutor, and the defense lawyers all suspected that both men had raped her before they killed her and that her rape-murder was opportunistic. That is, they had not planned to rape anyone when they first decided to steal a car but took advantage of the situation that developed.

In an attempt to bargain for her life and that of her unborn baby, Laura Foss told them that she would write them a $20,000 check so that they would have enough money and never have to commit robbery again. Bobby said that he agreed to take the check but that Johnson would not. He did not want to give his name to the woman so that she could write it on the check because he feared being identified. To reassure him, Laura wrote the check out to "John Doe" and handed it to Bobby. Unfortunately, her desperate attempt to trick them did not work. Johnson took the check from Bobby, tore it, and threw it to the ground. Laura was stripped, beaten, raped, and then shot.

The two men were convicted of abduction and murder and both received the death sentence. Bobby Moore's execution was only a few days away when one of his lawyers was watching Diane Sawyer on the *CBS Morning News*. Diane Sawyer was interviewing Dr. Dorothy Lewis about the carefully performed studies of violent delinquents she had recently published.[1]

The defense lawyer contacted Dr. Lewis and asked if she and her coworkers could examine Bobby Moore and several other convicts whose executions were imminent. The lawyer believed none of the inmates had been examined carefully, neurologically or psychiatrically. Dr. Lewis agreed to examine several men and asked me to join her. As a member of her team, I traveled to the prison and evaluated Bobby Moore and several other prisoners on death row.

Bobby Moore was a thirty-nine-year-old, large, well-muscled man. At first glance he did not look different from any ordinary man his age except that he was dressed in the red jumpsuit provided for the denizens of death row. I introduced myself to him and extended my hand in a polite greeting. He provided a limp handshake in response and returned my greeting, but I could not understand what he said. His articulation was indistinct. My difficulty was not just the result of his rural Southern accent but instead was caused by his dysarthria, a neurologically based incoordination of the muscles involved in the production of speech.

My communication with him was impaired further by his inability to answer simple questions simply. He was indirect and would go off on tangents, speaking of matters irrelevant to the questions I had posed. The deficiency in our communication extended to his use of pronouns. It was difficult to follow the subject about which he was speaking. For example, he indicated that he had been seriously ill as a child and had missed school for a prolonged period. Asked for what illness, he responded, enigmatically, "Pus in your skin."

The brain controls behavior, thought, and speech. When these functions are disturbed, the chances are that the brain is abnormal. Irrelevancies and idiosyncratic uses of language can indicate a thought disorder such as schizophrenia, some other psychosis, mental retardation, or a hearing deficit. However, cultural factors can also play a large role in aberrant behavior and abnormal speech. I had to assess these different possibilities during my examination of Bobby.

In my standard neurological examination of Bobby, I detected no hearing problem. His memory was poor chiefly because he had trouble concentrating. I asked him to repeat "leather belt, rubber ball, and blue." It took several repetitions before he mastered this task. He was unable to read beyond second grade level, and he could not perform simple arithmetic. Yet he was not retarded in every area. His fund of information was fairly good. He named the current president and his two immediate predecessors and could easily name five cities in the United States. His attention deficit could have been the result of psychosis and/or frontal lobe dysfunction. Indeed, there were several physical signs that pointed toward frontal lobe abnormality, such as discontinuous, jerky eye movements during visual tracking.

His head circumference, sixty-one centimeters, was greater than the normal range for an adult male. This deviation can reflect developmental abnormalities such as hydrocephalus in infancy. Hydrocephalus ("water on the brain") results from the inability of the body to absorb cerebrospinal fluid[2] as rapidly as it is produced. Often children and adults with hydrocephalus have residual motor and cognitive problems even when the condition spontaneously arrests and the pressure is no longer elevated. A computerized tomography (CT) scan or a magnetic resonance imaging (MRI) scan of the brain should be performed in a person with cognitive and motor problems whose head circumference is abnormally large. Neither test had been performed on Bobby Moore, nor had either test ever been discussed by the doctors who had seen him before me.

There were also lateralizing signs, that is, evidence of dysfunction on one side of his brain. There was a marked discrepancy between the rapidity and accuracy with which he could tap the tip of his left index finger against the crease of the left thumb as compared with the right. The same incoordination of the left side of his body emerged during the performance of other tasks that require fine motor coordination. The voluntary movements of the left side of the body are initiated by the right side of the cerebral cortex and are modified by the cerebellum.[3] The dysarthria and the motor deficiencies that I found in his left hand and leg might have reflected dysfunction on the right side of the cortex of his brain, or in the cerebellum. The cortex of the brain, the gray matter that overlies the entire surface, provides all cognition, behavior, and thought. The fact that his behavior and thought were also disturbed made it more likely that all his clinical neurological abnormalities were the result of cerebral cortical, not cerebellar, dysfunction.

On the day of my visit, psychological testing was also done. Bobby Moore achieved a verbal IQ score of 77 and a performance score of 85, for a full IQ of 80. This is low-normal, not retarded. Both the Bender and the Halstead-Reitan Batteries[4] indicated brain damage, especially in the right hemisphere but also bilaterally. The Rorschach[5] indicated psychosis. An EEG was diffusely abnormal, indicating that the electrical activity on the top of both sides of the brain was abnormal.

As first related by him, Bobby Moore's history was like filtered soup. Most of the lumps that give it character and flavor had been removed. He said that he was one of sixteen children fathered by two different men. There was violence in the household, and he had often seen his parents fighting, mostly verbally, with each other. Sometimes there was actual physical conflict. In fact, he recalled one incident in which his father aimed a shotgun at his mother. His older brothers and sisters intervened to save her. This sounded like a fairly dramatic event, but he provided no details and recounted the incidents without showing any emotion. He said that he had not been beaten "badly" when he was little. Specifically, he said he had been beaten for bad behavior by his mother with a switch that raised welts all over his body, but other siblings had been similarly beaten, and he nodded when I asked him if he thought these punishments were fair and not excessive. This history suggested abuse but, as I learned later, he had "cleaned up" the extent of his abuse and minimized and sanitized the intensity and severity of intrafamilial strife.

He said that he got into fights at school because, "People trying to think for me, trying to tell me what to do." He perceived teachers' requests of him as attempts to control him. He related multiple incidents when he believed that he had been "set up to be killed" by various people. He said, "You always have someone who dislikes you. It's a mean world, a jealous world, an evil world, but Jesus came." This brief statement may have reflected paranoia, idiosyncrasy, unusual logic, and religious preoccupation.

Many of the fights in which he had been involved were his response to hallucinations, misconceptions, or misinterpretations, all of which had a frightening and threatening aspect. Bobby and his mother had some very odd beliefs. He said that his mother had recruited him to help her cast voodoo spells on his family and neighbors. Bobby said he had told his mother about "visions" that he had seen. His mother interpreted these as visitations from the dead, and regarded them as ordinary, certainly nothing requiring a medical consultation. He would reach out to touch the dead people he saw, but his hand went through them. He heard the voices of men and women, whom no one else could hear, some of whom he could identify and some not. They spoke to him and to one another. He began to have these hallucinatory experiences at about the time he started elementary school and they continued until the time I saw him. As I was speaking with him, he claimed to hear these voices speaking at low volume in the back of his mind.

Thus, the results of the tests of brain function were concordant with one another. The neurologic examination, psychological testing, and EEG all indicated brain damage. Both the neurologic examination and the psychological testing especially implicated the right hemisphere and found evidence of psychosis. The history suggested psychosis and hinted at abuse. Bobby Moore's possible motivation to dissemble made it imperative to have other sources of information about his background and medical history. Nonetheless, Bobby had to be the starting point of the endeavor.

A judge granted a temporary stay of Bobby Moore's execution to allow further investigation of possibly mitigating factors. An investigator was hired who was able to interview many of Bobby's siblings and the minister who knew the family when Bobby Moore was a child. Without the efforts of this investigator, I might have concluded that Bobby had not been seriously abused.

The investigator interviewed Jimmy, one of Bobby's half brothers, who was the first son of Bobby's mother by her first husband. (Bobby Moore's father was his mother's second husband.) Jimmy said that he had left his mother's home after his father's death when he was thirteen years old because he "was tired of the beatings" he received from his alcoholic mother. Jimmy described how his mother would get drunk and would strip him naked and make him get into a sack, tied at his neck. She would then hoist the sack up toward the rafters of the shack in which they lived so that it would swing in the air. Then she would beat Jimmy with whatever instruments were available, such as broom handles, a shovel, pipes, and extension cords. His father occasionally protected him from the worst of the beatings but his father worked long hours. If his father heard about especially vicious beatings, he would beat Bobby's mother in retaliation. Then, she would beat Jimmy in retaliation for that. Beatings and violence were a major mode of intrafamilial interaction. Jimmy returned to his maternal home when he was seventeen. By that time, his mother had married her second husband, Mr. Moore, who also was an alcoholic. This couple conceived many more children, one of whom was Bobby Moore.

Another of Bobby Moore's older brothers, Tom, described his very early memories of his mother's drinking and recalled that she and his father would "hang around in bars a lot" and his mother would accuse his father of "laying out with other women." Tom said that his parents fought all the time. He often heard them arguing and saw his father striking his mother and vice versa. As a small child, Tom witnessed his father choke his mother into unconsciousness and beat her. Tom said that their fights were not limited to hand-to-hand combat but often involved guns. His mother kept a pistol with her all the time and his father had a loaded shotgun within easy reach in their home. They fought with every weapon they could get their hands on, including knives, belts, sticks, and guns. Sometimes their fights continued, intermittently, for day and night. The only reason they did not kill each other was that the older children would intervene at critical moments and would keep them apart.

The mother, at one point, was openly sleeping with other men and this was a major focus of familial fighting. Tommy, age nine, begged his father to leave home because he was afraid that his parents were both going to die in a fight. When he told his father this, his father burst into tears, packed his things, and left.

One of Bobby's sisters, Bessie, told the investigator how she had hated the weekends when she was little because on those days both her parents would get drunker than usual on homemade whiskey and her mother would come after the children with her gun. Their mother's wrath seemed to be especially directed toward her sons, and toward Bobby more than the others.

Bessie, like her older brother, Jimmy, described how the mother would beat all the children, requiring them to strip and "become stark naked" before she would whip them. One of her favorite sayings was, "I bought the clothes and the clothes didn't do anything." Every time they were beaten, they were beaten while naked. Many times the beatings caused blisters and bleeding.

The mother beat her children with plaited tree branches, a rope that had been tied in knots and soaked in water to make it stiff, an iron cord, or anything else she could get her hands on. She would often force her children to obtain the switch or rod she used for beating them. The children had to read their mother's mind to predict what instrument their mother wanted and had to provide one that was big enough to cause sufficient pain or the beating would be worse.

At bedtime, she often tied the children's hands (or legs) with ropes, one rope on each hand. The free ends of the rope were fastened to the bedposts, spread-eagling the children. In the morning, before the children woke up, the mother would remove the free ends from the bedposts and loop them over the rafter above their beds near the ceiling. Then she awakened them by hoisting them up into the air by their arms or legs. She would beat them while they were thus suspended. Such beatings and others, more conventionally contrived, were delivered before she went to work or they to school, seemingly unrelated to the children's behavior. Their mother said that they would "get it again and even worse" if they misbehaved during the day.

She gave neighbors permission to come into the home and punish the children whenever she was away from home. This license was eagerly pursued by some of them, according to Bessie and some of Bobby Moore's other siblings who were interviewed by the investigator.

Marla, another of Bobby Moore's sisters, said that she had witnessed her mother tying one of the boys to a tree and lighting a fire under him. He was freed by some of the older children. Her mother cruelly abused Marla as well. When she was about thirteen, Marla was forced to strip naked during her menstrual period. Her mother bound her hands and suspended her by her hands from a rafter and beat her in front of the mother's boyfriend. Marla told the investigator she felt "so embarrassed and ashamed."

All the siblings interviewed by the investigator said that they were expected to work in the fields. Their wages helped their mother to keep a roof over their heads, but there was never enough to eat. Many days and nights the children were hungry, even though there was food in the house. Their mother hoarded the food away for the famine that she expected when starvation would again loom before her as it did in the Depression. For this reason, she denied her children access to food and beat them if she thought they had eaten too much. Bessie said that the little ones had stomach cramps that she attributed to desperate hunger.

"To see food and not be able to eat was so confusing to the little ones. I begged for food in order to feed them. If it had not been for the generosity of some of our neighbors we would have starved," recalled Bessie. Bobby was physically large and had a big appetite. As "a big eater," he was beaten by his mother and siblings for consuming more than his "fair" portion of food.

All the siblings who were interviewed agreed that Bobby was the focus of the worst beatings because of his very evident vulnerabilities. He was "an odd one." Everything had to be told to him more than once and his mother thought that she could "beat sense" into him. His brother Everett described his mother tying Bobby to a bed, naked, and beating him "until the blood came. Most of his beatings caused bleeding."

One of the neighbors who was given permission to beat the children when the mother was away "gave the worst of it to Bobby." When he behaved badly this neighbor tried to break him by beating, and then would add to his terror by forcing him to crawl under his bed in the dark, insisting that he remain there until his mother came home. This confinement in the dark was also the practice of his mother who would threaten him with a pistol if he did not remain under the bed. She kept him confined under his bed in the dark for hours on end. According to Bessie, Bobby was "terrified of the dark, and being under that bed (in a dark room) with the door closed really affected him. I can hear him now, whimpering and pleading to get out. He remained afraid of the dark long after he became an adult." Another reason his mother and the neighbor would force him to stay under the bed was because they thought he was hyperactive. "That's the place they could make him go where they knew he couldn't get into any sort of mischief."

Bessie described how, as an infant, Bobby Moore had cried uncontrollably and how she had tried to comfort him without success. As he grew older, she realized that something was wrong with him. It took him longer to learn to walk than the others, and he had great difficulty forming his words. When he did learn to talk, he stuttered so badly that it was difficult to understand him. His school performance confirmed her suspicion that he was retarded. He was never able to learn to read or write and, eventually, the teachers gave up on him and passed him from one grade to another.

Bessie said, "He needed help and everybody in town knew it by the time he was school age. Our minister came to see my mama and told her that he would run into some problems when he got older if he did not get some help, but my mother ignored the minister's warnings as well as the warnings of several other people. She never acknowledged that he was not normal."

Several siblings told the investigator that Bobby would sit around the home, talking to himself, withdrawing and staring during parental fights. He would often climb a tree when stressed. On one occasion, he fell out of the tree, struck his head on the ground and was knocked unconscious for an hour. No medical attention was sought. Several of his sisters felt that Bobby was worse, mentally, after this fall but could not be very specific. They all agreed that he spent hours sitting in a tree. Even as a grown man, he would climb a tree and remain there for hours when he was upset. Because of this peculiarity and many others, adults and children, including his own brothers and sisters, cruelly teased and abused him. One older brother actually admitted to trying to shoot Bobby in retaliation for him wetting the bed in which they both slept. Bobby Moore wet the bed almost every night until his teens. For this he was beaten and teased.

His brother Charlie said that "everyone" referred to him as crazy. Few people easily understood his speech and most could not carry on a proper conversation with him. He was ignored by most people. He would sit in a room full of people, talking and carrying on a conversation with himself. When he did talk to others, his conversation was ignored. People cut him off or moved away and avoided him. He had trouble staying on one subject and could not keep his mind on one thing.

He participated with some of his siblings in acts of cruelty directed at small animals—drowning or shooting cats and setting them on fire. Cruelty to animals, fire setting, and enuresis (bed-wetting), supposedly predictors of juveniles who will become violent, are hallmarks of juveniles who are being abused.

At age sixteen, Bobby Moore quit school and sought a job. Bessie said that he was a hard worker and earned a good paycheck but gave all his money away. Because their mother never said that she loved any of the children or showed any kindness to them, all were starved for affection. Bobby's way of obtaining friendship, according to Bessie, was to give his money to anyone who paid attention to him. He was gullible and easily influenced, a potential problem some of his siblings warned him about to no avail. "He had friends for the first time in his life, even though he had to buy them."

Bobby Moore was beaten, starved, and humiliated consistently during his formative years. His brain was physically damaged early in life, prenatally through maternal drinking and postnatally by the constant stress of his abusive home. The development of his concepts of love, devotion, the value of life, and fairness was crippled. His paranoid thinking may have resulted in part from his life experiences and from physical and mental illness. He was described as "odd." In fact, he was mentally ill, probably schizophrenic. He talked to himself. His mind wandered. He had delusions and hallucinations. His mother demonstrated similar symptoms with her talk of witches and possession. Her unreasonable fear of hunger and her demonic, sadistic treatment of her children may have also been evidence of psychosis. It is likely that psychosis was genetically transmitted to Bobby Moore.

Bobby's social skills and repertoire for dealing with life's difficulties were severely limited. In fact, his coping mechanism was to climb a tree to get away from the social problems of his home.

He nursed feelings of rage that were engendered by his childhood treatment. His behavioral model was that of an angry, abusive, drunken, weapon-carrying mother or father who lashed out in unreasoning fury and terrorized the household. He was in mortal fear of his mother. Yet his unwillingness to criticize her to me and his willingness to share with her the hallucinations that she interpreted for him implies some degree of admiration. She was a powerful figure. Everyone cowered during her rages and beatings. He may have aspired to be like her, to control others, to make *them* cower, whimper, plead, and beg, to rise from the level of victim to the level of perpetrator. To this end, he was cruel to animals when he was a child, drowning cats in sacks or setting them on fire with lighter fluid.

Most of his siblings were able to avoid becoming violent criminals even though they too were abused. For this reason, it seems unlikely that abuse alone was a sufficient explanation for Bobby Moore's violent acts. Two of Bobby's siblings were in prison for homicide but thirteen were not. Why were thirteen of his sixteen siblings not murderers in that abusive household?

Most people who have been abused do not become abusers or violent criminals.[6] According to my theory, the ability to overcome the antisocial urges and impulses that abuse generates requires a nervous system that is functioning without intrinsic deficits. This would be reflected by normal neurological and psychiatric profiles. It takes a very good, solid brain, one that is unimpaired by neurologic or psychiatric disease, to overcome the tendency to violence that is engendered by consistent, long-term abuse delivered by parents or parent substitutes at a tender age.

Many psychotic and brain-damaged people live quiet lives. Criminality and violence are definitely not obligatory or even common concomitants of brain damage with or without psychosis. It seems likely that it was abuse interacting with brain damage and psychosis that sealed Bobby Moore's fate. In that interaction lay the awful seeds of destruction that so horribly took the life of a young, pregnant wife and a policeman who was performing his duty.

Bobby was easily led and desperately wanted someone to be nice to him. He would do anything for anyone who seemed nice to him, even give over his paycheck, according to his sister. He could not resist an invitation from a "friend" to participate in an armed robbery.

It is likely that none of the steps of the robbery that demonstrated forethought was initiated by Bobby Moore—stealing a car so Johnson's would not be traced, going to a grocery store to obtain a car, driving to a lonely spot in order not to be seen, killing

Laura Foss to avoid identification, and tearing up the check with which Laura Foss sought to save her life, having correctly judged that it would not help them. The same could be said for stalking the convenience store and waiting for customers to leave. The killing of the policeman and driving away were spur-of-the-moment-responses, goal directed but not planned, and not considered well in advance of the action. Only the gun was provided by Bobby and carrying a gun was his usual practice, not necessarily part of a plan. He was a follower, but he could also be vicious.

A rational person, thinking about a criminal act, assumes that the criminal is as rational as he and tries to find rational reasons that explain the criminal's modus operandi. A rational criminal, motivated by the desire to rob, might well have chosen the most vulnerable victim in a parking lot, a pregnant woman like Laura Foss. He might have killed her, stolen her car, robbed the store, in a premeditated fashion, and then tried to escape, killing the policeman in the process.

This scenario was proffered by the prosecution to explain the actions of Bobby Moore and Thurmon Johnson. Several facts were inconsistent with this simple theory. No money was taken. If the purpose of the exercise was robbery, why was nothing taken? More importantly, Laura Foss was not just assassinated. She was also beaten and raped. Why beaten? Why raped? These elements cannot be explained by the desire to rob and to eliminate a witness. The rape involved the bruising of a nipple, a punch on the chin, bruises on her neck and shoulders. She was pleading for her life, and it is very unlikely that she did anything to provoke such anger and hatred. Certainly she did nothing that could have justified or explained the criminals' actions against her.

Why then, did Bobby attack her? When Bobby had a cowering, pleading woman begging for her life at his feet, I imagine he was flushed with the conqueror's pride. He was a man who had helplessly lain bruised and bleeding, starving under his bed in the dark, whimpering to be released, while tortured by visions of ghosts and totally controlled by his armed mother, whose gun was in his hand as he hovered over the captive woman. He was not able to marshal his social inhibitions, to suppress his impulses, to conform his behavior to the requirements of law, or to release his captive unharmed. He could not surrender the power he suddenly was able to exercise. His lack of conscience, lack of empathy with his victim, and inability to feel the dimensions of the tragedies he created on that day can be labeled as moral insanity, sociopathic personality, or antisocial personality disorder. Abuse, brain damage, and psychiatric illness were the interacting causes of his behavior, whatever its label.

Conscience, empathy, and their partners, morality and ethics, are not inborn qualities. They are learned in childhood. In other words, the neural networks that underlie morality and ethics are established by means of environmental influences during the development of the brain in childhood. Experiences that lead toward morality can mold the development of conscience only if they exist in the environment and if the child has the neural capacity to benefit from them. Parents are primarily responsible for teaching these concepts to their children, but additional seminal experiences are derived from other family members, neighbors, teachers, and clergy. Bobby Moore may not have been able to derive any benefit from whatever wholesome influences he may have encountered because of his brain damage and psychosis, although there were not many positive environmental influences.

Bobby's childhood "lessons" prevented him from developing his conscience and a sense of remorse. He was awakened by being hoisted toward the ceiling with a rope and was beaten, with the promise of more if he was "bad." However, he could not help being "bad." He was damaged: hyperactive, of limited intelligence, and psychotic. His speech was abnormal in its mechanical expression and did not make sense to those who made the effort to understand him. His social skills were minimal. Like many children with attention deficit disorder in abusive families, he was a magnet for excessive punishment. The lesson? No matter what he did, he was powerless to escape beatings, and he could not behave in a manner that would not result in a beating or confinement under his bed in the dark.

His life was actually in danger in his own home. His mother was fully capable of shooting him or of burning him alive. His world was an evil place; only power counted. Power came from guns, sex, and strength. Punishment was the only method used in his home to affect the behavior of others. To escape the horrors of his world, he either dominated or withdrew. Given the opportunity to dominate, he committed rape and murder; at other times he climbed a tree.

Bobby Moore's acts of murder and rape can be viewed as disinhibited, impulsive, and the result of poor judgment. Whatever planning he did in committing his crimes was fragmentary and poorly conceived, although there were predatory elements. All these features can result from frontal damage. The neural networks contained in the frontal lobes inhibit primitive urges and instinctual behavior. Accordingly, damage to parts of the frontal network can result in the loss of inhibitions.

The effect of alcohol intoxication mimics and aggravates the effects of frontal lobe damage. Under the influence of alcohol, the executive frontal network ineffectively censors the drives and instincts even of normal people. The effects of alcohol are even more pronounced in brain-damaged individuals. Bobby had been drinking just before the first homicide.

Yet his vicious acts cannot be attributed solely to alcohol. The vast majority of intoxicated individuals are not violent. Most alcoholics are not criminals, let alone violent criminals. But more than half of all homicides are committed by people who are under the influence of alcohol.[7] Violent alcoholics are excessively suspicious (paranoid) and have been physically abused in childhood as compared with nonviolent alcoholics. Alcohol consumption correlates with homicide even more than mental illness.[8]

The psychosocial history is a critical dimension in the production of violence even when the evidence of structural neurologic damage is overwhelming. For example, there is the famous case of Charles Whitman who climbed the tower at the University of Texas in Austin and randomly shot and killed twenty-three people before he was shot and killed by the police. The medical examiner discovered a malignant tumor located deep in the brain during his autopsy, the presence of which had been unsuspected during his life.

Was the brain tumor the cause of the mass killing? There are thousands of people each year with deeply located, malignant brain tumors who have never behaved in an antisocial manner, let alone killed anyone. It is crucial to note that Whitman came from a broken, dysfunctional home. His perfectionist father beat him, his mother, and his brothers. In interviews following the shootings, Whitman's father declared that he didn't think he beat Whitman's mother "any more than the average man" beat his wife, and if he ever had to do it over, he would not have been "so easy on the kids." As he told the *Austin-American Statesman* in August 1966. "I don't think I spanked my children enough."

Abuse may have differentiated Whitman from others with similar brain tumors. Possibly the tumor damaged his capacity to control the violent impulses that were engendered by his childhood environment. Attributing his murderous behavior solely to the brain tumor would be as much a gross oversimplification as attributing his behavior solely to abuse.

In the case of Bobby Moore, brain damage and alcohol abuse were factors in his poor planning and inability to control his impulses, but the origin of his criminal impulses were likely the result of abuse. His vicious acts were the product of his early environment and the effects of brain damage. Though it surely was a crucial factor, brain damage alone was an insufficient explanation for what Bobby Moore did.

Violence rarely can be the result of paranoia alone. The determining aspects of some homicides are psychotic delusions of persecution. The psychotic individual acts violently to protect himself from what he imagines to be malign forces around him. Command hallucinations (auditory experiences of voices telling the person to do something such as kill himself or someone else) can accompany paranoid delusions. Though Bobby Moore was psychotic, his killings did not appear to be his response to paranoid delusions or command hallucinations. Though Bobby Moore was naively trusting of unsavory and exploitative "friends," he was excessively suspicious, too. His inability to trust people, especially women (teachers, sisters, and neighbors), and his anger, intensified by psychosis, may have increased the likelihood that he would attack Laura Foss and the officer, and it may have contributed to the lethality of the attacks.

This story of Bobby Moore's early life, obtained by the investigator from the siblings and minister of Bobby Moore, moved the judge. He reduced Bobby's penalty to a life sentence without the possibility of parole. The judge's decision to spare Bobby's life was based on the findings that Bobby had been horrendously abused from birth to sixteen years of age, as well as his psychotic condition and brain damage. As a clinician, I was moved by the story but puzzled too. I thought it very likely that the abuse he had suffered was causally related to his commission of two vile and odious murders, but at that time I had not formulated my current concept of how abuse can lead to murder.

CHAPTER 4

Genes, Geography, and the Generation of Violence

There are so many theories that concern the cause of violence. Almost everyone has an idea and many approaches contain an element of truth. Researchers in fields as different as molecular biology and social science have tried to identify the cause of violent behavior. Like the blind men who each examined a different part of the elephant and who accurately described one part of the beast, the concept of the whole animal eludes them. Each ascribes violence to the one thing in which he is interested and ignores the evidence of other factors, yet each contribution is a speck on a mound of other evidence.

"Isn't it all genetic?" I have been asked that question many times. If violence were genetic, we could leave all the explanations for it to biologists and geneticists.

The theory that an abnormal gene causes human violent criminal behavior has been contradicted by clinical adoption studies. In adoption studies, the criminal histories of adopted males were compared to the criminal histories of both their biological and their adoptive fathers. These studies found that genetic influences were not significant in causing violent crime.[1] Variation in societal rates of violence from generation to generation, from social group to social group, and country to country as well as fluctuations in the rates of violence within the same population would be inexplicable if violence were simply genetic.

The theory that a tendency to violence is inherited could be correct in a limited sense. Many serious mental illnesses like schizophrenia, bipolar affective disorder (manic-depressive psychosis), and recurrent depression are inherited. All of these genetic mental diseases can cause paranoid delusions and auditory hallucinations when they are severe, which, in combination with abuse and neurological deficits, can lead to violence. Bobby Moore was psychotic, probably schizophrenic, and so was his mother, but I do not think that the genetic component of Bobby Moore's behavior would have been dangerous to society had it not been associated with abuse and neurological deficits.

In the 1960s, Richard Speck's behavior supported the theory of genetic blight as a cause of violence. He entered the Midwestern apartment of a group of student nurses and methodically slaughtered all but one who escaped by pretending to be dead. She identified him. After his arrest, he was studied and found to have a chromosomal defect. Each male is supposed to have one Y chromosome and one X chromosome. Speck had two Ys and one X. The "Speck" theory was that males with an extra Y chromosome were predisposed to criminal behavior. To test out this theory, surveys were done of violent prisoners. An unexpectedly high prevalence of XYY was found. This seemed to provide an explanation for the higher prevalence of violence among males. It was all genetics. Too much Y was bad. It provided an extra measure of maleness, of violence.

Later work found little support for this hypothesis. When the general population was surveyed and XYY was uncovered, scientists found no correlation of the chromosomal deficit with violent behavior.[2] The original studies had contained a classical statistical sampling error that was so obvious in retrospect that it was embarrassing. Of course all the men they had originally identified as XYY were violent. They had only tested violent men. Even the men who were not XYY in that sample were violent too.

Mental retardation is one of the most common features of patients with chromosomal abnormalities of all different kinds. Trisomy 21 (mongolism) is an example of this. The mental retardation associated with XYY is not as severe as in Trisomy 21, but mild mental retardation is prevalent in males with XYY. Richard Speck was mildly mentally retarded.

Perhaps via mental retardation, the extra Y chromosome in his case provided his neurologic vulnerability that would have never been realized were it not for other factors about which nothing was said at the time, such as if Speck had been abused in his childhood. Was Speck mentally ill? paranoid?

Two biological theories[3] have commanded much attention within the scientific community: one is neurochemical, the other relies on molecular genetics. The neurochemical theory is that low brain levels of the transmitter serotonin cause violence. Serotonin is an important chemical messenger in the brain. It is released by certain nerve cells when they are stimulated and acts on other cells, usually inhibiting them. For about twenty-five years, biological researchers have reported a correlation between a low cerebrospinal fluid concentration of serotonin and various behavioral abnormalities, including violent and criminal behavior. Although more than one hundred studies have been published on this subject, more recent studies have not really confirmed the results of the earlier ones.[4] At first, the behavior characteristically associated with low concentrations of serotonin was found to be depression. Later studies correlated low serotonin not with depression but with aggressive behavior. Subsequent studies correlated low serotonin levels with impulsive aggressive behavior and with alcoholism. In sum, the later studies do not replicate the earlier ones. Not every individual with low serotonin levels is depressed, aggressive, impulsive, or alcoholic, nor does every individual with any or all of these have low serotonin levels.

Even if a statistical association were to be established between low serotonin concentration in the brain and some well-defined violent or criminal behavior, causation would not be proven by this link. Which comes first? Could both be caused by something else? Do low concentrations of serotonin cause the abnormal behavior or does the abnormal behavior trigger physiological responses in the body that lower the serotonin concentration? Could abuse lower serotonin levels and, through some other mechanism, also cause depression? Conceivably, low concentrations of serotonin may predispose affected individuals to violence by inducing depression. It is certainly true that depression triggers violent behavior in the predisposed. Yet the predisposition to violence is also determined by abuse, paranoia, and brain damage. According to the theory I have proposed, violence is the product of interacting factors. One of these may indeed be related to low serotonin brain levels, but it certainly is doubtful that the level of serotonin fully explains or causes violent behavior by itself.

It has not been possible or desirable to perform a lumbar puncture to measure the serotonin concentration in the cerebrospinal fluid of Bobby Moore or in any other murderer I have examined. The status of research on the biological basis of violence is far too preliminary to permit any meaningful conclusions about the role of serotonin.

The other genetic approach that has received significant attention from the scientific community postulates that human violence is encoded in our genes and that a gene for violence may exist. Some people have said that if such a gene could be identified, it could be inactivated in utero.

Some years ago, the medical-scientific community was excited by a report in the journal *Science*. The authors found an association between an abnormal gene and aggressive behavior. They showed that the gene that makes monoamine oxidase type A (MAO-A) the enzyme that degrades serotonin, was abnormal in aggressive males from a large Dutch family who pushed and shoved one another and made inappropriate sexual advances to close female relatives. The mutation the aggressive males carried led to the absence of MAO-A. This report was the first that implicated a specific gene as being the cause of aggressive behavior toward others.[5] Although the defect in the gene for MAO-A might well have been responsible for some of the mental and social symptoms of affected male members of the Dutch family, there is little prospect that this condition is an important cause of violent criminality.

The aggressive mannerisms of affected males in the Dutch family did not actually reach criminal levels in most of the males who carried the genetic abnormality, although some tendency in that direction did appear. The defect in the gene for monoamine oxidase-A is extremely rare even in the criminal populations that have been screened for its presence. Therefore, this genetic defect is unlikely to be an important factor in more than a tiny proportion of violent criminals, even if the correlation of the defective gene with violent behavior in the affected families were to be confirmed. The aggressiveness shown by the affected males may have resulted from an interaction of neurological deficits with learning disability, paranoia, and the possibly abusive manner in which the affected children had been handled by their parents. The authors provided no information whatsoever about abuse and paranoia in the family and the family members were very secretive about their intrafamilial social relationships.[6] This makes me suspect abuse—both physical and sexual—in the family.

I do not wish to be too negative about the theory that genes can influence violence. There is certainly good reason to believe that there is a strong genetic disposition to violence in animals. Certainly, the great variation in aggressiveness among different animals, even within the same species and in different species, suggests this. Wild rats are more aggressive than albino rats; German shepherds than poodles; lions than Siamese cats. Even in some animals though, early handling makes a big difference. German shepherds can be sweet and loving, and cruelly handled poodles can be quite nasty. The role of nature and nurture in animals' behavior can be argued, but the epidemiologic studies in *humans* concerning this subject have failed to support a genetic cause of violent crime. Certain diseases of the brain that can precipitate violence have a genetic basis (schizophrenia, bipolar affective disorder, obsessive-compulsive disorder), but most patients with these conditions are not violent.

There have been two major independent sociological approaches to the understanding of violence. One sociological approach identifies hardship and deprivation as societal causes of violence.

Individuals are viewed as being driven to violence by their unfortunately low location in the social spectrum. This socioeconomic approach directs attention to poverty, poor housing, general economic inequality, and inequality based on ethnicity and race.

There is a maddening lack of precision to this sociological theory. Why should bad housing cause violence? Why was violence more prevalent in the 1980s than during the Depression years of the 1930s, when a third of the workforce was unemployed, people were poorly clothed and poorly housed, and starvation was common. Though I admit that violence is more prevalent among the socio-economically deprived, the most compelling argument against this theory as a comprehensive explanation of violence is that most members of society, even the most deprived, are not violent.

The other major sociological theory of violence emphasizes cultural orientation. This perspective depicts criminal violence as the outcome of normal processes of social learning. Certain groups allegedly endorse values that are supportive of violent behavior. Consequently, well-socialized individuals belonging to such groups are predisposed to employ violence as a means for dealing with common, interpersonal disputes. This theory has been used to explain the high rates of homicide and use of guns that have been widely observed for certain demographic groups in society, for example, the Mafia. Everything we know about the Mafia is anecdotal and far from the scrutiny of scientific analysis.

There are, however, variations in the rates of violence, geographically and regionally, that are statistically significant. The United States has one of the highest rates of homicide in the highly developed world at 7.4 per 100,00 in 1996, compared to a European average of 1.2 per 100,000 for the same year.[7] In a *New York Times* article, in which these facts were presented, the author, Fox Butterfield, discussed the large variation in the rates of violence within the United States. He pointed out that the former slaveholding states of the old Confederacy have the highest homicide rates in this country, led in 1996 by Louisiana with a rate of 17.5 murders per 100,000 people. The ten states with the lowest homicide rates are in New England and the northern Midwest, with South Dakota's the lowest at 1.2 per 100,000 people.[8]

Richard Nisbett, a professor of psychology at the University of Michigan, analyzed the homicide rates for white males.[9] Nisbett found that in medium-size and small cities, the homicide rate of southern white males was two to three times higher than rates for white males in the rest of the country. In rural areas, the southern rate was four times higher. This is a socially significant fact, and it is not likely that climate has much to do with it. After all, Finland has one of the highest homicide rates in Western Europe and one of the coldest climates.

There is likely to be a cultural difference between the South and the rest of the country that might provide a permissive attitude toward violent expression. In his book *All God's Children* (Knopf 1995), Butterfield proposes that slavery and segregation fostered in the South a disinclination to prize human dignity and feelings of self-worth because society was rigidly stratified largely on racial grounds. The people at the bottom were not supposed to be worthy. Instead, an elaborate system of demonstration of respect for others developed.

The warm expressions of welcome toward strangers that are so common in the South, and so appreciated, may be manifestations of this "respectfulness." Children are taught to say "Yes, sir" and "No, ma'am" out of respect. Respect is something others provide. A person without a sense of self-worth who is deprived of respect can become dangerous. "Dissing" means disrespecting. Disrespect has often been claimed by criminals whom I have examined as the justification for homicide. Butterfield suggests that the respectfulness system has been carried by African Americans to the urban centers of the North and West.

Closely related to this may be the fiercely punitive attitude toward homicide in the South. Texas and Virginia together are currently responsible for about half the executions in the United States. Most of the remaining executions are performed in the South.

There is also a racial aspect to violence in the United States that can be seen from several perspectives. African Americans are more likely to be victims of violence. In 1987, the homicide death rate among all American men aged fifteen to twenty-four, according to the World Health Organization, was 22 per 100,000. By 1994, it had risen to 37 per 100,000. The homicide rate for young African-American males was 167 per 100,000 in 1993. In inner-city Philadelphia from 1987 to 1990, 40 percent of all young African-American men had suffered a violent assault serious enough to send them to a hospital emergency room. African Americans were six times more likely than whites to be murdered in 1998.[10]

If child abuse is a major cause of violence and African Americans are disproportionately the victims of violence, then child abuse should be more common among African Americans than among whites. And in fact, in 1996, the number of African-American victims of child abuse was almost twice that in the national child population. White children were a lower proportion of child-abuse victims than their representation in the population.[11] Child abuse is the fourth leading cause of death for all American children aged one to four, but second for African-American children of that age.

If the theory of violence that I have proposed is accurate, child abuse should also be more prevalent in the South than in other regions in the United States. This is also the case. Reported cases of child abuse are 50 percent higher in the southern states than in New England.[12] Of course, reported cases may be the tip of the iceberg. It is my (unconfirmed) impression that "whipping" children, especially boys, is normative practice in much of the South and among African Americans and is not considered abusive. This was illustrated in the case of Sam Wise.

Sam Wise was the murderer of three young people who were unlucky enough to work the late shift in a coffee shop in an elegant neighborhood of Baltimore. He shot each in the head with a single bullet. This act of cold mass murdering was the capstone of Wise's career of armed robberies that had included a previous killing and an attempted homicide. Wise's lawyer, Nancy Baskin, asked if I could find any mitigating factors. Wise was facing the death penalty. The entire community had been outraged by his crime that was unsolved for weeks and the subject of daily newspaper coverage. Wise was a black man who had killed three whites. Baskin was desperate to discover something that would save her client from death and get him a life sentence without parole. She told me clearly, "Sam Wise is intelligent, comes from a middle-class family, and was not abused." I suspected otherwise and arranged not only to examine Wise and review his records, but also to interview his mother. I wanted to know, from her perspective, if Sam had been abused.

The records and the exam were not surprising. Wise had a history of attention deficit hyperactivity disorder (ADHD), psychiatric hospitalizations for psychotic episodes with hallucinations, suicidal ideation, and a major depressive disorder. His neurological examination and psychological testing were abnormal. This was as I expected, but it was not only his lawyer who had said Wise had not been abused. Wise and his mother both agreed that he had not been abused. Wise said that his father had never physically punished him but had denied him television to discipline him. Later in the interview, Wise allowed that his mom had beaten him, but not more than once a month and always for good cause. The beatings were brief, minutes at most, and never drew blood. The story his mother told did not jibe with this description.

Evelyn Wise, Sam's mother, came to Baskin's office to meet with me. She was an African American of light complexion, about seventy years old, and dressed in her Sunday best with a beige suit and a small hat with a veil. This was clearly an important meeting in her estimation. Yet her eyes were downcast and her posture suggested guilt and shame, shoulders hunched, neck flexed. She spoke softly and often said, "Yes, sir" in addressing me.[13]

A college graduate, she had worked at a medical school for forty-two years as a laboratory technician in the animal room. Her husband had worked in the post office for decades before he died from the complications of diabetes and stomach ulcers.

She had always wanted children, but had none of her own, so she and her husband finally adopted Sam and his younger sister. Sam was hyperactive from day one. She enrolled him in nursery school when he was two years old.

"Yes, sir. If he misbehaved in nursery school, I'd discipline him," she said.

> My husband objected to this and would say to me: Are you trying to kill him? My husband demonstrated permissive love. I didn't like to argue with my husband because when he got upset, his stomach acted up and that was more work for me. I'd have to puree his food. He couldn't help with the housework and then I would have to do everything—shopping, cooking, cleaning—so I'd beat Sam away from home.
>
> He'd been told to do things he didn't want to do. He wouldn't come back in after playing in the yard. He fussed and hit other children. He lost his temper easily. He would not engage in organized activities. If he hit someone, I'd whip him.
>
> His fighting got worse in kindergarten. He didn't like to be teased and then he'd start to fight. I'd beat him except on the weekends when my husband was home. Then I'd let it go. It was not right. I should not have let it go . . . but I was so tired. I didn't always have the strength to beat him when he misbehaved.
>
> Then, when he was eight years old and I found out from the school that he was hyperactive and learning disabled, I had my first visit with a specialist and they started family visits. My husband could not accept that there was anything wrong with his son, so I had to deal with everything alone. Yes, sir, I did everything alone.
>
> The doctors said I shouldn't beat him so much and suggested alternate ways to punish him and prevent his bad behavior by keeping him busy in school. The doctor helped with suggestions, but her suggestions didn't work when I tried them, no, sir, they didn't work. So, I kept a collection of whips around the house—in the china closet, on the refrigerator, in the basement, in the car—so I would always have it ready to beat him when he misbehaved. I also used a belt at times.
>
> He was a head banger as a toddler and I had to beat him for that too. He was changeable, like a Dr. Jekyll and Mr. Hyde. He wasn't afraid of anything. He was a daredevil child.

With great feeling and downcast eyes she mused, "I was not as strong a disciplinarian as I should have been. I'd punish him for something and he'd do something similar a short time later." He didn't seem to remember.

She acknowledged that the beating did not work. "I'd have to call him two or three times to come in the house. I'd give him too long to comply. Then, I'd beat him and the next day it would be the same. He'd not come when I called. Sometimes, I'd just fuss at him but sometimes I'd whip him 'country style' ["Country style" is a colloquialism that means "severely with a switch."] or in the back seat of the car. He'd be crying and jumping around and real upset and then just two days later, he'd do something else he'd need to be beaten for. He didn't learn the lesson. His memory seemed a problem for him."

When she was really angry, the whipping lasted longer. It went on for minutes. She would make Sam say, 'I will not to do such-and-such any more.' She would give him one lick with each word. Then, he'd have to repeat it.

When I asked Mrs. Wise, "Considering everything you know now and what has happened, what would you have done differently in raising Sam?" Her head bent toward the floor, Mrs. Wise looked stricken, sad, and answered: "I should have been more strict. Yes, sir. I should have found the strength to beat him when he deserved it, and I was too tired."

I asked Mrs. Wise, "What about your father? Where were you raised?"

"I was born and raised in Arkansas. My father was very strict. He was a man of his word. You had to obey him right away. He was the one who helped us with our schoolwork."

"How nice," I said enthusiastically, glad to find a nugget of bright relief.

"Yes, sir. I suppose it was nice, but if we didn't catch on to the schoolwork, he'd punish us. He beat us. I didn't have trouble, but my sister had problems. He spanked her during math. To this day, my sister blanks out at any arithmetic due to fear, even getting change at a store. He was too strict. The fear of beating was pervasive. There was no second chance in my house. No, sir, no second chances."

This intelligent, well-spoken, middle-class grandmother recognized her father's brutality but did not see it in herself and castigated herself. She blamed herself, felt shame and guilt, for her slackness, for not having been more severe.

Whatever underlies the violence that characterizes the South, or any group or family in which violence is prevalent, we are entitled to ask: What differentiates the majority of individuals in that group who are not violent from the minority who are violent? If there are special vulnerabilities within violent individuals, these factors may be within their brains. Sam Wise was mentally ill and neurologically impaired. The beatings interacted with these to create a confluence of factors that caused murder.

Such a confluence of factors was certainly present in the case of Bobby Moore. All of his sixteen siblings were treated similarly but only three of the seventeen children became violent criminals. There must be some individual factors that differentiated those who became violent from those who were not violent. The social acceptance of violence, deprivation, and the effect of racism and poverty may contribute to violence in the South and among African Americans, but these are insufficient explanations for the relative rarity of violence. According to my concept, other vulnerabilities within the brains of the violent individuals, like neurological deficits and mental illness, explain violence in the minority of abused individuals who become murderers. Added to the changes in the brain wrought by abuse at an early age, these vulnerabilities can prove lethal.

The frequent and prolonged history of physical and sexual abuse committed by a parent or parent substitute has been pervasive and extreme among the 150 or so murderers I have seen. It has been the life experience of 94 percent of all the murderers I examined in a consecutive five-year period and reported in 1995.[14] Extreme abuse was also present in the histories of thirteen of the fourteen individuals Dr. Lewis and I examined who were on death row for homicides committed before they were eighteen years old and in fifteen condemned murderers we had examined just before scheduled executions.[15]

Paralleling the rise of homicide rates, the prevalence of physical, sexual abuse, and neglect doubled between 1986 and 1993 providing some explanation for the rise in violent crime rates. Possibly contributing to the greater prevalence of violent crime among males and among the poor, boys were 24 percent more likely than girls to suffer serious injury from abusive maltreatment. The poorest children with family incomes of less than $15,000 per year had more than twice the rate of physical abuse as children in families with annual incomes between $15,000 and $30,000 and almost twelve times the rate of abuse as those in families with incomes of more than $30,000. Significantly higher rates of abuse occurred in single parent families and those with more than four children. Birth parents were the perpetrators in 72 percent of the cases. Of these, about half were physically abused by mothers. Although the National Incidence Study[16] did not address the cause of child abuse, the report implied that the rising use of illicit drugs was strongly associated with abuse. Other studies support the finding that more severe parental physical abuse is directed toward boys.[17]

These population studies link sociological insights to my theory. Poverty, large families, single-parent families, and illicit drug use are associated with the likelihood of child abuse. Being a male child usually increases the severity of abuse. Child abuse is one of the main causes of violence, although not the only one, because most people who have been abused do not become violent. Nonetheless, child abuse is a major factor. The increase in the reported prevalence of abuse has preceded and paralleled the rise of violence in our society.

Children with ADHD (see chapter 6) are probably much more likely to be abused than normal children as their hyperactive behavior may be less acceptable to abusive parents. Abused children and adults are more likely to develop post-traumatic stress disorder (PTSD) if they are neurologically impaired.[18] Since ADHD is more common in boys, they may be magnets for abuse much more than are girls. Not only is ADHD more common in males but so are traumatic brain injury and other forms of brain damage. Brain-damaged children and those whose neurological and/or mental illness makes their behavior difficult for parents to deal with may also invite more parental abuse. In this way, ADHD, abuse, and PTSD are linked.[19] Reported child abuse in developmentally disabled children is 2.1 times as great for physical abuse and 1.8 times as great for sexual abuse. Of all children physically or sexually abused, 15 to 17 percent had disabilities. This is an understatement as it includes only reported abuse and excludes children in institutions.[20]

If brain damage and child abuse are causes of violence, their increasing occurrence in the United States, especially in certain vulnerable segments of our society, might well explain the rise of violence in the United States over the past three decades.[21] For example, twenty years ago, about 5 percent of white children were born to single mothers. Currently, the rate is more than fourfold higher. Twenty years ago, about 25 percent of African-American children were born to single mothers. Currently, two-thirds are. These figures reflect societal and familial disintegration. This disintegration probably underlies and contributes not just to the rising prevalence of substance abuse, teenage pregnancy, child abuse, and brain-damaged children but to the resultant vector of these destructive forces—violence.[22]

CHAPTER 5

Wrath: Repression and Release—The Effects of Frontal Lobotomy

Probably the shock of actually being in a prison and seeing condemned men for the first time in my life colored and made more vivid my memory of that occasion. The kind of abuse that Bobby Lee experienced was not different from that of many other murderers that I have examined since, but somehow his story stands out in my memory. The same is true of the next case. It illustrates the tremendous influence that frontal lobe damage has upon violent behavior. Dirk Donovan was one of the five condemned men whom I examined during my first visit to death row.[1]

"You don't have the guts to rob no 7-Eleven," said Gus Werlie to his coworker, Dirk Donovan, in the midwestern steel mill in which they were both employed.

"Yes, I do too," Donovan replied.

Donovan was a big man, only twenty-two, but very strong. He could lift and carry whole sheets of steel that, for safety, usually were carried by four men from one part of the mill to another. Donovan would do almost anything if dared and enjoyed showing off his physical strength, often risking life and limb in the process. He was among the youngest of the workers and enjoyed pleasing the older men, especially his boss, the owner of the mill, Bill Isaacson.

Donovan was no genius and certainly was not good-looking. The left half of his face was paralyzed and scarred, and he was blind in one eye from birth. Nonetheless, Isaacson was impressed by the young man's strength, energy, capacity to work hard, and his very evident desire to please his boss. It was obvious to everyone that he was Isaacson's favorite. When Isaacson praised him to his face, it was one of the great moments in Donovan's life. He beamed at Isaacson and would have given his life for him, had Isaacson asked him.

No fellow worker could criticize Isaacson in front of Donovan without risking a serious fight. Isaacson had stuck by him, too. Once, while working in the steel mill, Donovan mistakenly thought that Jim Cassidy, a fellow worker, had insulted him. The insult had been directed at a third individual, but Donovan initiated a fight of mortal intensity with Cassidy. Donovan was stabbed and Isaacson fired Cassidy for starting the fight.

Donovan's role as Isaacson's pet was galling to a few of the other workers. They could easily provoke Donovan by teasing him, but they had to do this very carefully. Once Donovan almost killed a coworker with his bare hands because he had been belittled. It took six men to pull Donovan off the other man.

Gus Werlie was being sly. "You couldn't rob no 7-Eleven because you'd have to kill the clerk." Donovan was puzzled. "Why would I have to kill him?" he asked. "Because he'd recognize you and report you to the police," said Werlie.

That night, Donovan entered a convenience store in a small town, armed with a knife. He was affable and engaged the clerk, Tom Morely, in some small talk. Morely reported that Donovan, in a very pleasant manner, had said that he had come that night to rob the store. Morely thought he was kidding but when Donovan took out his knife, Morely knew that he was in trouble. With a smile, Donovan told Morely that he was going to kill him so that Morely would not be able to identify him and have him arrested.

Morely was terrified but thought quickly. Maintaining a sincere, friendly but slightly hurt tone, he assured Donovan that he would never report the robbery to the police. He said in effect, "I don't care if you rob the store. It's not my money that you are stealing. Do you think I want to work here? I don't. The only reason I'm working here is because I'm wanted by the police, and it's the only place I could get some work without tipping off the cops.

If the police find me, they'll put me in jail. I escaped from jail. I'm just like you. I need money. I would never turn you in because if I did, they'd find out who I am and arrest me and send me away for a long time." Morely said that they should stick together.

Reassured by this profession of communality of interests and cares, Donovan took the meager contents of the cash register and departed. Soon after, he was arrested, having been identified by the shaken Morely who had called the police immediately after Donovan left the store. Donovan was charged with armed robbery, tried, and convicted. He was sentenced to prison for a relatively short time because it was his first major offense. Morely, the chief witness against him, commented on the disarming inappropriateness of Donovan's friendly manner.

To minimize Donovan's sentence, Isaacson provided a character reference. After two years' imprisonment, Donovan's job at the steel mill was restored by Isaacson. Donovan remained at that job for about a year until he came under the influence of a girl from Alabama who wanted to move back "home."

Donovan left his job and accompanied his girlfriend. Once in Alabama, he was unemployed and short of cash, unable to purchase for his girlfriend what she desired. Donovan decided to rob a convenience store at night. He would prove his manhood and cleverness and not make the mistake he had made before. While his girlfriend waited approvingly and expectantly outside in her car, Donovan held the thin, short clerk at knifepoint, took the small amount of cash available in the till, and told the clerk to precede him to the back room.

In reconstructing the crime from the autopsy, it seems that he plunged his knife into the young man's back with a fatal blow. Mortally wounded, the man spun around and offered a weak defense with his arms. In turn, Donovan stabbed him tens of times in the chest and arms and cut his throat.

Seventy-two hours after the murder, while his girlfriend waited in a nearby motel, Donovan called a used car dealer and asked him to stay late so that he could come to his lot to look over the cars. The car dealer obliged. Donovan approached the man, who was alone in his office, after hours.

A fierce struggle ensued. The victim, though overweight and elderly, put up a valiant defense. The office was destroyed; a window frame was broken; furniture and equiptment were scattered. Donovan stabbed the victim twenty times. The medical examiner said that many of the wounds were defensive, but any of five wounds could have been fatal. Another measure of the savage nature of the killing was the deep slash wound of the throat that almost decapitated the victim. The amount of money Donovan stole from the convenience store clerk and the automobile dealer totaled about $100. An investigation led to his arrest, and he confessed to both murders.

His motive for the crimes seemed clear to the police and the prosecutor. He wanted money, hence the robberies. The homicides were his means of preventing his identification. With his first experience with armed robbery as his paradigm, he would not make the mistake of leaving his victim alive again. In fact, before meeting with a court-appointed lawyer, Donovan himself had provided this explanation to the police in his confession. He was tried and found guilty of the first murder and was sentenced to life imprisonment.

He was in jail, awaiting trial for the second murder. This time his life was at stake as the prosecutor was trying to obtain the death penalty. Donovan decided that he should try to eliminate a witness who could incriminate him—his brother-in-law, Daryl. Donovan used a pay telephone clearly marked "monitored." In a taped conversation with his mother he not only again admitted committing the two murders but, heedless of the danger to himself, tried to convince his mother to arrange to have Daryl killed.

The reason he wanted to kill Daryl was that Donovan had confessed to Daryl that he had committed the murders. Indeed, he had good reason to expect that Daryl would testify against him, but Daryl was only one of several people to whom he had confessed. These included the police and his mother. Of course, his girlfriend knew all about the murders, too. Most of these testified against him. His desire to have Daryl killed counted heavily against him at his sentencing trial as an aggravating factor. He was convicted and was sentenced to death for the second murder.

Curiously, after the trials, he steadfastly denied having committed either crime, confessing to the police, ever calling his mother to arrange for Daryl's murder, or having told others of his involvement in the murders. What a stupid, series of transparent lies this seemed to be! It appeared to represent a foolish attempt to avoid responsibility. He suggested, blandly, to his frustrated lawyer that another person might have both committed the crimes and confessed.

In summary, Donovan had displayed several glaring errors in judgment. He had committed three robberies and two murders, ostensibly for money, but realized almost no monetary gain from any of them. The first robbery was the result of a dare! He spoke to the police and his brother-in-law about his role in these crimes without first obtaining the advice of his lawyer. He tried to arrange to have his brother-in-law, Daryl, killed to prevent adverse testimony, even though Daryl's death would not have helped Donovan's case at all. In so doing, Donovan used a monitored phone.

Despite the overwhelming evidence that Donovan's judgment was impaired, his own court-appointed attorneys and the prosecuting attorneys accepted without question his own original confession and his account of his motives for the murders. No one had thought to question his state of mind, let alone the condition of his brain. He had not been carefully assessed for the presence of neurological or psychiatric disease, though everyone who met him knew that his judgment was impaired.

Here I faced a dilemma quite similar to that posed by Louis Culpepper, the sexual molester: the common tendency to conceive of moral and ethical judgments as coming from some source other than the brain. Where do morality and ethics reside? In the spirit, wherein lies the divine spark? Volition, as commonly conceived, derives from the mind, an entity that is separate from the brain. Even if this view about will is correct, ethics, morals, volition, and the mind all operate through the brain's activity, and they can be interrupted by physical injury.

This analogy may help: The only way in which a composer can present to an audience what he has conceived is by having his musical composition performed by an orchestra. In a similar way, the expression of a person's will must involve the brain. The brain is the symphony orchestra of a person's spirit.

There are at least three explanations for a dreadful musical performance: the composer did a bad job of composing, or the orchestra did a bad job of playing what the composer had written, or both. If an investigation revealed that the orchestra had played using broken instruments, it would be likely that instrument failure was at least partly responsible for the poor performance. If investigation of a miscreant reveals that his brain is broken, it is likely that brain failure was at least partly responsible for his unacceptable behavior.

The dilemma of assigning responsibility for behavior in brain-damaged individuals is very difficult to resolve. I think that is why the case of Phineas Gage is remembered.[2] It hit a nerve. The famous story of Phineas Gage exemplifies the kind of deficit that frontal lobe injury produces. Gage was a hard-working, reliable, moral, ethical foreman who supervised a crew of men laying railroad tracks in Maine about 150 years ago. Gage set up the gunpowder used for blasting. He was injured when a premature explosion blew an iron rod through his frontal lobes. He recovered from the injury and could walk, talk, read, write, calculate, and remember; however, he was a changed man. He had become immoral and unethical. He began to drink, swear, fight, and carouse. He became a derelict. By means of a traumatic injury to the frontal lobes, he suddenly developed elements of what today would be called antisocial personality disorder.

All thinking is done in the brain. If judgment is impaired, the brain is probably impaired. That is a reasonable hypothesis, but what evidence supports it? As it turned out, there was abundant reason to suspect that Donovan's brain was damaged.

A new set of lawyers handled Donovan's case in appeal of the death sentence. They initiated a preliminary investigation of his background and found some features that they thought might be relevant but could not be sure how. Birth records showed that Donovan's delivery had been traumatic. Forceps had been used and, as a result of their misapplication, Donovan had been blind in his right eye from birth. Traumatic deliveries such as this are often associated with brain damage.

Donovan's school performance was poor in the primary grades. He had what today would be called attention deficit disorder, learning disabilities, and a conduct disorder. He could not pay attention in class and was delayed in the acquisition of basic skills like reading, writing, and arithmetic. He was impulsive and aggressive and required frequent disciplining for fighting. He did not get along well with other children in his school classes, but he was always physically bigger than the others and his strength enabled him to protect himself. He was described as taking pride in his ability to win fights. He had no friends, though he was anxious to please his teachers, and aside from his distractibility, slow learning, and frequent fights with other children, he was good-natured in class. He had been made to repeat grades twice and left school permanently as soon as he could, at sixteen, in order to work.

He was a prodigious manual laborer, strong, able, and willing. It was easy for others to take advantage of him. Any fellow laborer who was nice to him could get him to do his work, egg him on to foolhardy feats of daring, and could even borrow money and not repay it. But Donovan had one very weak point: if he were teased or belittled, he would instantly turn vicious.

He worked hard at a variety of jobs while still at school and gathered enough money to buy a car. He was reckless and had a "heavy foot" by his own account. At age sixteen, he had a head-on collision as he tried to pass another vehicle at high speed on a blind curve. Donovan was seriously injured and was rushed, unconscious, to a hospital. A tracheotomy was performed and he was placed in the intensive care unit. He had no recollection either of the accident or of the first two weeks he spent in the intensive care unit. He sustained a number of fractures. His left foot was badly damaged, and the right side of his face was permanently paralyzed as a result of the crash. He was kept in the hospital for several weeks.

His family said that after this accident his behavior had deteriorated. He became very difficult to live with. Because of his altered behavior, he had been thrown out of the house by his father and was living on his own before his seventeenth birthday, but the details of the interaction within his family at this time were vague.

The traumatic delivery, with resultant blindness of one eye, and his poor school performance pointed toward brain damage. The motor vehicle accident seemed likely to have further damaged his brain. The accident had caused significant traumatic brain injury as evidenced by prolonged unconsciousness and amnesia for the events that both preceded and followed it. His personality seemed to have been changed by the accident. His face was paralyzed. These were historical features that cried out for a neurological assessment. There were other features of his history that should have triggered a psychiatric and social evaluation.

Dirk Donovan was not his name. It was the name he liked to be called. His official name and the one on his birth certificate was Vincent Zepolino. He did not like that name because it was the one his father had given him, and he did not like his father. He did not believe that Zepolino, his mother's husband, was his biological father.

His mother had always vehemently denied Donovan's suspicions and insisted that he was Zepolino's natural son. Not believing her or knowing who his real father was, Donovan had, years earlier, made up the name he preferred to be called.

His mother even sent a devastatingly cruel letter to Donovan, filled with four-letter words and other crude invectives to describe him, that he received a few days before the date on which he was to be executed. In this, possibly the last communication he would ever have from his mother, his mother insisted that Zepolino was

Donovan's real father and that Donovan was a stupid, motherfucking piece of shit not to acknowledge that.[3]

Donovan's lawyer told me that he did not understand how Donovan could have been emotionally unaffected by that farewell letter from his mother, yet he gave no evidence of having any emotional response to the letter. He talked about it without difficulty, blandly, a faint smile on his face.

This was the more remarkable because his mother told one of Donovan's lawyers, in the strictest confidence, that Donovan was correct. Zepolino was not Donovan's biological father. She had become pregnant before she married Zepolino and gave birth to Donovan. After she married Zepolino, he adopted her baby boy (Donovan) and then she went on to have several more children with Zepolino.

A perfunctory psychiatric assessment of Donovan had concluded that he simply had "antisocial personality disorder." That he had behaved antisocially was not in doubt. The issue was why. The useless tautology that equates antisocial acts with antisocial personality is the diagnostic redoubt of clinicians who do not wish to take the time to perform a thorough assessment or to think about the issue. Almost all of the symptoms associated with antisocial personality disorder can also be manifestations of frontal damage (impulsiveness, poor work history, aggressiveness, recklessness, little ability to foresee consequences, limited insight, limited remorse, indifference, lack of empathy, irresponsiblility). This list of symptoms sounds like a description of Phineas Gage or Dirk Donovan.

A psychologist had performed an IQ test. It was normal. Donovan had expressed no remorse for what he had done. He even denied having killed the two victims. Therefore, reasoned the psychologist, there was no brain damage, no mental illness; it was a simple case of antisocial personality disorder.

Diagnostic labels and understanding are often very different. One of my friends in college suffered from a chronic itching around his anus. He had consulted many physicians, none of whom could diagnose or treat the condition. When one doctor said, "Why, you have *pruritis ani*," my friend was immensely relieved and said, "At last, someone understands my condition. He has made a diagnosis. Now it can be treated!" My friend's satisfaction was short-lived. He checked the word *pruritis* in the dictionary: it means "itching." The Latin title *pruritis ani* provided no more understanding or treatment than the English ("itching of the anus"). The diagnosis "antisocial personality disorder (APD)" is the *pruritis ani* of violent behavior.[4]

An individual who stabs a man dozens of times during a robbery may be motivated by a colossal emotion that has little to do with the apparent goal of the crime (money). It also implies defective impulse control. The commission of two similar murders, characterized by vicious overkill, for a total of less than $100 indicates terrible judgment, poor planning, and defective prioritizing. What would cause such emotion, lack of control, and poor judgment?

On my neurological examination, it was obvious that he was damaged. He limped. The right side of his face was paralyzed. His right eye was blind and smaller than the left. It had been so since birth. He carried some impressive facial scars and the scar of a previous tracheotomy that dated from the motor vehicle accident when he was sixteen.

The right side of the brain controls movements on the left side of the body. Therefore, the fact that Mr. Donovan had considerable weakness of the left leg, that he limped and could not hop on that leg, was consistent with the existence of right-sided brain damage. Yet that foot had been badly fractured in the accident. Could his limp have derived from an orthopedic rather than a neurologic cause? The reflexes settled the matter. There was a left Babinski sign and also a right Babinski sign.[5] Thus, there was evidence of damage to both sides of his brain.

He also demonstrated a snout reflex; discontinuous, jerky eye movements during visual pursuit; and an inability to perform reciprocal, alternating motor acts with his hands. These are all signs of frontal lobe dysfunction.

His conversation was almost as obviously abnormal as his limp. He maintained an inappropriate joviality and talked excessively in good-humored fashion. He seemed utterly oblivious to the fact that he was a condemned prisoner whose execution was only a few days away. The clerk in the convenience store whom Donovan had threatened in his first robbery had noted a similar inappropriateness of affect. His uncaring attitude and inability to feel the seriousness of his situation were typical of the thought processes of individuals with frontal lobe injuries.

He had trouble answering a simple question simply. His answers rambled and inconsistently focused on the subject under consideration. He was also irritable, according to the history, and misinterpreted social situations in a paranoid manner. The stories of his fights in the steel mill attested to this.

He said with pride that he had never lost a fight. The most reliable source of his feelings of self-worth seemed to derive from his ability to beat up others. When he became a teenager, he began to carry a knife. When I asked why this was necessary, at first he laughed and said he did not know. With further questioning, he indicated that this was for protection and that he frequently felt threatened without having any particular focus of concern. The carrying of a knife, therefore, was an adaptation to his paranoid thoughts.

Irritability was a longstanding personality trait. Combined with paranoid thoughts it could be dangerous. Once, before the accident, he had become so angry at his sister that his father had to subdue him lest he kill her. His irritability and "low boiling point" changed so markedly for the worse following the accident that his parents had to expel him from their house, leaving him, at age seventeen, to fend for himself.

According to Donovan, this turn of events did not entirely displease his father, Zepolino, who, it seems, never really treated him as though he were truly his biological son. They had an adversarial relationship. He said that his father treated him like a stepchild, making him do much of the menial work for the family. He said his father criticized him unfairly and never brought him presents or candy when he returned from trips, though he did bring gifts for the younger siblings. This was why Donovan hated the name "Zepolino." Frankly, this seemed to be an inadequate basis for the pervasively negative feeling Blake expressed for his father.

My prior experience with violent delinquents led me to strongly suspect that Donovan had been severely physically abused as a child but no such history was forthcoming from him. Zepolino, Donovan said, beat him with a switch when he misbehaved. Although Zepolino never showed love or affection for Donovan when he was a child, Donovan denied that Zepolino had badly beaten or abused him either. Rarely, Donovan said, Zepolino "had slapped around" his mother but neither of his parents had ever abused him.

Family secrets are often associated with malignant events. The secret concerning his paternity was a major one that raised a red flag for social pathology. It was impossible to gather adequate information directly from Donovan or his mother on this subject. Donovan did not provide a very clear idea about why he did not like his father, and his mother denied any serious intrafamily conflicts. This rosy view was not credible. After all, Donovan's brother-in-law testified against him, Donovan had once nearly killed his sister and wanted to kill his brother-in-law, Donovan's parents had expelled him from their home, his father had "slapped around" his mother, and there was the paternity issue. His farewell letter from his mother was not loving, to say the least. This was not a big happy family, but what was wrong?

Because evidence relevant to mitigation had not been introduced at the first sentencing trial, the judge ordered a second. In preparation for it, the judge permitted us to complete our clinical evaluation. At my request, he ordered neuropsychological tests, an MRI scan of the brain, and an EEG, and he permitted the defense lawyers to engage the services of an investigator to review the social background of the defendant.

These additional efforts confirmed and extended my clinical findings and conclusion that Donovan had traumatic damage to his frontal lobes. The neuropsychologist found evidence of damage to both frontal lobes, using the Wisconsin Card Sort Test,

Trail Making Tests A and B, a continuous performance test, and the Category Test from the Halstead-Reitan battery.[6] The EEG was abnormal. The MRI scan of the brain showed bilateral frontal white matter lesions, of traumatic origin, that effectively represented bifrontal leukotomies and severed the connection of the frontal gray matter to the rest of the brain.

I was pleased by these results because they confirmed my clinical diagnosis. Even if the MRI and EEG had been normal, I would have been convinced that my findings were significant, but in this case, all the tests were concordant.[7]

The social investigator traced a half sibling of the defendant, Frank Zepolino, to the midwestern city from which Donovan had come, about 1,500 miles away. At the time the investigator was able to contact him, Frank's children had just been removed from his home by court order. He had been sexually molesting them and caused them to sexually molest each other. The investigator was able to review the reports of the social workers that had led to the removal of Frank Zepolino's children from his home. These reports recorded some of the activities within the family during Donovan's childhood and indicated the kind of home in which he and his sexually abusive half brother had been raised.

From age six or seven and continuing for almost a decade, Donovan had been used by his mother for her own sexual gratification. The younger children were often locked out of the house when the father was absent in order that Donovan could perform cunnilingus on his mother. Other sexual games were played as well and Donovan was forced to engage in sexual displays and activities with his younger siblings for her edification. His mother gave him license to beat his younger siblings to force their participation in orgies. His mother rewarded him with something approaching kindness in return for sexually satisfying her and punished him by ridicule and beating.

Zepolino was a brutal man who hated his stepson, probably in part because he was the son of a rival and partly because he was, in fact, a rival himself, though of tender age. The beatings he delivered to Donovan were frequent, severe, prolonged, and accompanied by ridicule and humiliation. On one occasion when he was about nine years old, Donovan climbed a tree to escape a beating at the hand of his stepfather. In order to reach him, Mr. Zepolino chopped down the tree with an ax.

Without this information from Frank Zepolino and his social service records, we would never have known the extent of Donovan's physical or sexual abuse. Child Protective Services had been called by the neighbors, but their efforts were ineffective.

Like many other individuals who have been horrendously abused as children, Donovan had completely repressed these experiences. Even after being confronted with the evidence, he had no recollection of these painful scenes.

Sometimes, when abuse is relentless and intolerable, children totally block it out in order to survive and experience it as though it were happening to someone else. Later on, they may have absolutely no recollection of the abusive events. This is called dissociation and is part of what we now call post-traumatic stress disorder. The different names (Donovan, Zepolino), the "forgotten" murders and confession, and the bizarre use of a monitored telephone for a conversation he later denied having suggests a form of dissociative disorder once called multiple personality disorder. When I first evaluated Donovan, I had no idea of how completely he could have repressed the ghastly experiences documented in his half brother's records. The revelation that dreadful abuse had occurred helped to explain why the impulses unleashed by Donovan's brain injuries were so rage-filled and murderous.

In view of this new data, I attempted to reconstruct the probable dynamics of his crime. Abuse underlay his violent impulses. His only sources of pride were physical strength and its use to intimidate others. The abuse he had experienced as a child diminished any moral compunction that might have prevented his use of physical force. His mother actually encouraged him to impose his will on others through violence.

He was easily manipulated by anyone who offered him friendship and was much in the power of a woman who directed him to obtain money for her benefit. He planned the robberies and probably deliberately, exuberantly, and in premeditated fashion intended to murder. He probably enjoyed demonstrating his physical superiority to his girlfriend. The meager reward (less than $100) and the potential cost (his life) were not factors in his planning.

I think that he sincerely did not remember the murders, but whether or not he did, a similar dynamic has been present in most of the serial murderers I have evaluated. True, Donovan "only" killed two and because of the stupidity of his behavior that reflected the severity of his brain damage, he was more easily caught. There is little doubt that he would have committed a third murder at another time, probably in the same way, and thus entered the fellowship of serial murderers (three or more victims). Most of the serial killers I have seen have been brain-damaged but not as severely as he was. Grotesque physical and sexual abuse are other constant elements in serial murderers.

Brain damage seemed to have interfered with his ability to plan properly. He seemed to be unable to anticipate dangerous circumstances, to postpone or discriminate in achieving his goal. His goals were inappropriate. Anxiety and fear reactions were dampened by brain damage, but he was left with explosive, unmodulated anger that resulted from minimal stimulation. He was especially provoked by being teased and called "stupid." His brain damage created a constant encephalopathy that was very much like that of a man who is constantly drunk.

Were it not for frontal lobe damage, he might have been able to control the violent impulses engendered by abuse because he would have been able to foresee and emotionally respond to their consequences. Were it not for the abuse, there would have been no violent impulses to control. It was abuse that endowed the frontal brain damage with dire significance and frontal damage that invested the abuse with the license to kill.

I have often been unable to determine exactly when and how a violent person sustained brain damage. Donovan's history provided clear answers to these questions. Most of the individuals I have seen on death row have been brain damaged in their very earliest years, many in utero. Often the mothers of death row inmates are alcoholics or drug users and have used these drugs during their pregnancies. Some subjects, like the young Cynthia Williams (chapter 2), have shown the stigmata of fetal alcohol effect. Many mothers gave birth during their early teenage years when their motivation to obtain counseling and adequate medical care was absent. Prenatal care, good dietary habits, the use of vitamin supplements were all lacking, thus setting the stage for brain damage in their offspring.

Exposure to lead-based paints in substandard housing is common. One study indicated that 17 percent of African-American children have detectable lead levels in their blood.[8] Toddlers who are neglected and hungry can ingest the flakes of leaded paint. There is no safe blood level of lead and the developing brain is especially susceptible to its toxic effect.

Traumatic brain injury from shaking, beating, motor vehicle accidents, and fighting are other common causes of brain damage. The frontal lobes and executive functioning seem to be the most sensitive to the deleterious effects of all these causes of brain damage in young children.

Like Donovan, most of the violent individuals I have seen on death row had or should have had a preceding diagnosis of attention deficit hyperactivity disorder with conduct disturbance in childhood. Some had been treated with methylphenidate (Ritalin). Those not so diagnosed were not necessarily normal. Some lived in areas remote from standard medical care and did not have access to this diagnosis or any therapy.

Proper diagnosis and treatment were often lacking for various reasons. The wanderlust, alcoholism, or social fragmentation of the parents often made it impossible to provide a sustained relationship with a particular physician, health worker, or teacher. Missed appointments and incomplete records are common. Some parents had odd ideas about the medical care of children, preferred to save money, or for other reasons would not see or consult with medical personnel. Paranoid mistrust of physicians and government clinics has been a common reason for avoiding care.

This springs from psychiatric illness in the parents combined with fear that the authorities will discover that the parents have been abusing the children.

Still others believe that the frequent, painful beatings that they deliver almost daily to their children do not constitute abuse. School and medical authorities are reluctant to intervene and sometimes share the abusing parents' view that beating is an acceptable parental prerogative. Some school authorities continue to practice corporal punishment today. Medical authorities often wish to avoid the prolonged, expensive legal battles that parents might initiate against them and avoid any mention of abuse; for they are obligated by law in most jurisdictions to report instances of probable abuse. Children are terrified and will not speak of abuse, lest their "tattling" become known to their parents. The children have reason to fear for their lives. Those who grow up this way may perpetuate fear in others through their violent acts. The behavioral hallmark of the previolent child is usually subsumed under the rubric "Attention Deficit Hyperactivity Disorder" and it may be worthwhile to consider this further.

CHAPTER 6

Immaturity, Mania, Mistreatment, and Miscreancy

What would one expect the behavior of a primary-school child to be like if his frontal lobes were not working normally? Since the frontal lobes help to focus attention, prioritize, and exclude the extraneous, the child would have a short attention span and would respond to irrelevant stimuli. He might be hyperactive. He would be very likely physically awkward (as the supplementary motor area is in the frontal lobe) and socially awkward, given to displays of temper and impatience. Yet, he could have a relatively normal IQ that he was not fully using, in the words of an old schoolmaster, "not working to his full potential."

The child would find it difficult to assimilate the basic scholastic skills like reading, writing, and arithmetic at an age-appropriate level. Even if he had the capacity to acquire these skills, he might be delayed in achieving competence in them because of his short attention span. Normal IQ and the ultimate development of reading, writing, and arithmetic skills implies normal posterior brain functioning. Insufficient attention span and delay in the rate of acquiring skills in the face of a normal IQ implies frontal dysfunction.

The child might lack social graces and appear immature. He might say and do things that others would not wish to hear or see. The child might not be able to determine the effect of his actions and speech on others, and he might not care about the adverse impressions he was creating. These are the symptoms of attention deficit hyperactivity disorder and of frontal impairment.

The high frequency with which I have encountered the symptoms of attention deficit hyperactivity disorder (ADHD) and conduct disorder (CD) in the childhood records of murderers has increased my interest in those conditions, though it is obvious that most children with ADHD do not become violent. CD tends to be the result of abuse. ADHD, on the other hand, is probably the result of poorly functioning frontal lobes or regions with which they connect, evidenced by neuropsychological testing, MRI studies, EEG, SPECT, and PET scans.[1]

ADHD is a behavioral syndrome, a complex web of related symptoms. As I conceive it, ADHD is not a sickness in itself, but a feature of other illnesses. Its connection to the brain makes it the proper concern of neurologists, psychiatrists, and psychologists. It is a syndrome that has different causes, treatments that vary depending on the cause, and a prognosis that depends on the causes. Like headaches and epilepsy, treatment and prognosis depend heavily on the cause of dysfunction.

Non–socially disabling ADHD is seldom permanent and most children outgrow it. For this reason, the symptoms of ADHD may meld with normal development, and the diagnostic criteria for ADHD are vague. As many as 5 percent of all primary-school children may have an attentional disorder. Most likely this is related to variation in the rate of development of the frontal lobes and the mylineation of its fiber processes. The children whose frontal lobes are "hooked up" with the rest of the brain earliest are "very mature for their age." The ones whose frontal connections are slowest to develop have "ADHD."

Myelin is a fatty insulating material that surrounds nerve fibers. Without myelin, fibers do not carry impulses, and incompletely myelinated frontal lobes of children only gradually join the grid of electrical activity that operates the brain. "Immaturity" and "as yet unmyelinated" are related concepts. The characteristics of normal immature behavior (implusive, inconsiderate, risk-taking, careless, irresponsible, inconsistent and variable, moody, and childish) are consistent with those of an individual whose frontal lobes have not yet been fully myelinated.

In contrast to brains that are slow to develop (but within the broad range of normal development), damaged brains may never become normal. Behavioral problems and learning disorders may persist into adolescence and beyond.[2] It should be an embarrassment to the fields of pediatric neurology, pediatric psychiatry, and child psychology that these specialties have not developed or else do not use the instruments for the assessment of frontal lobe/executive function in children of primary-school age who have ADHD. Though ADHD is one of the most common reasons for referral of a child to a pediatric neurologist, most pediatric neurologists do not assess such children properly. They check the head circumference, the functioning of the lower parts of the nervous system—the brainstem, cerebellum, and spinal cord—but often ignore the cerebral cortex.

No normative data *exist* for school-age children with regard to the signs of frontal dysfunction that are so helpful in adults (see the appendix). It is important to have normative data for children as much that is normal in very young children would be abnormal in adults. Infants normally display reflexes that are abnormal in adults, and normal young children cannot perform complicated repetitive motor tasks that normal adults can perform. At what age do primitive reflexes and the inability to perform complex motor tasks become reliable indicators of brain pathology? Pediatric neurology has no answer.

Instead of doing their own tests to assess the function of the cerebrum, pediatric neurologists and child psychiatrists refer children for psychological testing. Children take IQ tests, achievement tests, and sometimes projective tests like the Rorschach. All these tests primarily assess the functioning of the posterior two-thirds of the brain. None of them are sensitive measures of executive or frontal functions. It is rare for a child neurologist or psychologist to test ADHD with the most appropriate tools. The Wisconsin Card Sorting Test and Trailmaking A and B have been standardized down to age six and are very indicative of executive dysfunction. Child neurologists and child psychiatrists are not trained to perform these tests and seldom specifically request them. They order "psychometric testing"—a vague term. Unless the referring doctors specify the tests he/she wishes to have done or the functions he/she wishes to be tested, the psychologist is left to determine this alone. It would be like an internist ordering blood tests. No competent internist would order "blood tests" without specifying which ones because he does not know what tests to order or what functions they assess.

Often children with ADHD have minor abnormalities of coordination on neurological examination. They can be clumsy physically as well as socially. They may act as if they were retarded, but their IQs are characteristically normal—low normal, but still normal. Sometimes there is a marked discrepancy between ability areas on the IQ test, but IQ testing seldom provides the answer to the questions, "What is wrong with this child? Why does the child have ADHD?" If the behavior is abnormal and the psychological testing is normal, it is likely that the wrong test was done. Normal tests of liver function do not mean the patient does not have kidney failure. Testing the back of the brain does not rule out failure of the front. The following story illustrates several of the different diagnostic considerations of ADHD.

Mrs. Bloehm, a woman in her late forties, was sitting in my windowless examining room, telling me in a quiet, gentle voice about the behavioral difficulties of her twelve-year-old son, Stanley. It was not necessary for her to say much, for while she spoke, her son, seated at her side, made loud noises and ejaculations. He did not scream, but his volume exceeded that necessary for normal conversation in the room. Sometimes his sounds were inchoate, and sometimes they were comments on what his mother was saying.

Stanley remained in constant motion, standing at times and waving his hands, fanning himself, as if to say, "It's very hot in here." He turned off the overhead light at one point, thrusting us into total darkness. His mother responded to this by addressing him in an unemotional tone, "Please turn on the light." He complied after a three-to-five second delay. He seemed in good spirits and smiled and laughed while he behaved inappropriately.

The visit was driven by desperation. Mrs. Bloehm and her husband were afraid that their son would kill someone during one of his rages. Until then, he had never really hurt anyone or come close to hurting anyone other than his sister, but he was growing so rapidly that she and her husband had almost lost physical control of the boy. For the past six years, when he had a tantrum, they had been able to restrain him during his rages only by immobilizing him in a bear hug. After fifteen to twenty minutes, Stanley returned to relative calm, and his parents would be able to release him.

They did not beat him. His father had spanked him once or twice at an earlier age, but the results of this were not felicitous. Stanley seemed to be propelled by spanking to new heights of unacceptable behavior.

In nursery school and kindergarten, his behavior was not outstandingly different from the other children. He had many invitations to the birthday parties of classmates, and he had two or three friends with whom he socialized regularly. In first grade this changed. He began to lose his temper easily, and his manner of verbal and physical expression became unacceptable. Changes in his routine unsettled him. During angry rampages he broke toys, dishes, and household furnishings while screaming and threatening at full volume. He did this at home, mostly, Mrs. Bloehm told me, but he had little opportunity to act badly in the homes of others. After one or two rages in the homes of friends, there were no further invitations from the other parents.

Stanley's parents attributed the change that they noted in his behavior when he was six to the birth of Melissa, Stanley's sister. Indeed, they had to take special precautions lest he seriously injure her, like locking her nursery door and never leaving Stanley unattended when he was in the same room as Melissa.

Diagnosed with ADHD, Stanley had been treated with Ritalin and other stimulants. Low doses of Mellaril (an antipsychotic neuroleptic) and very small doses of Elavil (an antidepressant) were without any benefit. Ritalin actually made his behavior worse.

Now that Melissa was five and Stanley twelve, Stanley's threat to her had become even more serious. "He often hits Melissa," recounted Mrs. Bloehm, "or throws things at her, and he would really injure her if my husband and I did not intervene. Anytime he perceives what he believes to be preferential treatment of his sister, he is likely to have a tantrum and to attack her. When he is in the midst of a tantrum, we cannot reason with him, but often his perception that his sister has been given something that he has been denied is simply wrong."

As Mrs. Bloehm spoke of Stanley's hitting his sister, Stanley jumped to his feet. He pounded his left palm with his right fist as a baseball player might strike his glove. As Mrs. Bloehm spoke of his throwing things, Stanley stood and pretended to throw objects. He did this with a big smile and chuckles and said, "I'll kill her. I'll smash her—that little . . . !"

Mrs. Bloehm said that although Stanley was an immediate threat to his sister, he had not until a few months earlier posed a danger to his parents. A few recent incidents had almost turned ugly with relation to her, and Mrs. Bloehm was concerned about the escalation.

As she was telling me these things, Stanley suddenly turned toward his mother. In a flash he removed her glasses from her nose and, laughing, placed them upside down on his own nose. His mother responded calmly in an even, soft, matter-of-fact voice that contained no portent or tone of threat or pleading and asked him to return her glasses. The spectacles fell from his nose to the (gratefully) thickly carpeted examining room floor. Chuckling and smiling, he raised his foot in the air as if to stomp on the glasses. I was alarmed.

"Don't do that!" I demanded. Perversely, my injunction seemed to prolong his charade. He did not smash his mother's glasses. After a little while he picked up the glasses from the floor and, teasingly, tried to hang them from a coat hook on the wall.

Mrs. Bloehm repeated calmly, "Give the glasses back to me, Stanley," and he complied. I was very much impressed by her cool, calm manner. She never once shouted, threatened, or struck him, though I needed every ounce of self-control to follow her lead. His actions were clearly unacceptable and yet he had struck no one in my office, did not break his mother's glasses, and had not stomped on them.

The Bloehms had taken Stanley to many psychiatrists, psychologists, and counselors whose desultory efforts had been unsuccessful. The psychiatrist who referred Stanley to me, Martin Stein, is one of the most talented and intelligent physicians with whom I have had the pleasure of working. After sensitively obtaining the history, he asks himself, "From where in the brain are these symptoms arising? What disease could be causing that part of the brain to malfunction? What treatment might work?" This is the proper diagnostic-therapeutic approach in neurology and should be the same in psychiatry. Dr. Stein typically treats difficult patients with success who have not been helped by other doctors.

Dr. Stein had already determined that Stanley's EEG was normal. A brain MRI scan was also normal. When I examined Stanley neurologically, I found many "frontal signs," but there are no published norms for these tests for twelve-year-old boys. Clearly the boy was not functioning like a normal adult, but I could not determine from the physical examination if he were damaged or just immature and within the broad range of normal for a twelve-year-old.

His schoolwork was mixed. He was very articulate with a good vocabulary and he did well in science, but he did poorly in mathematics and spelling, and his reading skills were far below average. This was not so much the result of an incapacity to learn to read or to calculate. In fact his reading and arithmetic skills had gradually advanced over the years and were continuing to improve. It was just that his poor attention span limited the rate at which he could acquire reading and mathematic skills, and this consigned him to the bottom of his class.

His IQ was superior; the full-scale score was 112, but there was a twenty-point spread between verbal and performance scores. The psychologist said that this suggested "organicity"—in other words, something neurologic. Though his academic record was not good, it was not failing. The chief problem area was his behavior, the tantrums and rages. These lasted for twenty to thirty minutes, occurred two or three times a day at home and at least once a day at school, and posed a threat to the safety of his classmates and teachers as well as his sister. Because of his behavior, he had been expelled from three different schools and had been placed in several different programs in those schools in the years between first and sixth grade.

Whatever instability that had occurred in Stanley's family seemed the result, not the cause, of Stanley's behavior. Stanley's home environment was absolutely nonabusive. When either of Stanley's parents could stand his misbehavior no longer, each would provide a break for the other. Mrs. Bloehm seemed to possess the greater reserves of patience, but there had been several close calls with regard to her physical safety during Stanley's tantrums in the six months before I examined Stanley.

Within two years Stanley would be fourteen. As a teenager with access to illegal drugs and the wrong friends, the Bloehms realistically felt that Stanley might destroy his life. He was hovering on the brink of destruction: crime, arrest, even homicide, and incarceration.

The family history is an important diagnostic tool in the evaluation of mental disorders, as the most serious illnesses (schizophrenia, mania, depression, Tourette's, obsessive-compulsive disorder) often run in families. Mr. Bloehm's family had members who were gamblers, alcoholics, and drug abusers. Stanley's paternal grandmother had suffered from depression, and her older brother, Stanley's great-uncle, was an alcoholic who had committed suicide. This history suggested bipolar affective disorder, though Stanley's father, Mr. Bloehm, had never shown any evidence of mental illness. He was a college graduate with a master's degree. He was a stable, effective person, according to his wife.

What was wrong with Stanley? His behavior seemed to me to be out of the normal range, even though I have no neat criteria for defining what the limits of normal twelve-year-old behavior are. He could have been among the lower 5 percent of normal twelve-year-old children who are very immature. It seemed to me that he was too active, too excited, too excitable, and too dangerous to be considered within the normal range.

He also could have sustained brain damage. True, the history did not suggest this. His mother's pregnancy, his birth, delivery, and development were normal. Still, unidentifiable congenital factors can cause abnormalities of brain development without a known history of what caused the damage or what part of the brain is abnormal. The physical examination cannot and did not differentiate normally *un*developed from *ab*normally developed brain. The normal EEG and MRI scans identified no abnormality, but neither test provides a definitive endorsement of normality. The psychological tests had indicated that there might be something neurologically wrong by the wide point spread between his verbal and performance scores, but this was not definitive. No psychological tests of executive function had been done. His frontal lobes, 40 percent of his brain, had not been assessed psychologically. The school psychologists and private psychologists who had tested Stanley had not used tests of executive function. The neuropsychologist would not test children.

Ritalin, a stimulant drug that increases attention span, had worsened his behavior. I counted this fact as indicating that his problem was not developmental. Ritalin helps most children with developmental ADHD. Ritalin makes mania worse.

There was depression, gambling, alcoholism, and suicide in Stanley's father's family. Could Stanley's behavior have been the result of mental illness? Certainly, Stanley was not depressed. He smiled, chuckled, and enjoyed his unacceptable behavior. He may have had moments of disappointment and upset to which he could not adjust successfully, but depression did not seem a likely cause of his ADHD syndrome. The unsuccessful trial of ineffectively low doses of an antidepressant was inconclusive.

Stanley showed elements of childhood autism: learning disorder, lack of friends, and peculiar gesturing (like flapping his hands). However, he had been normal until age six, and autism usually begins in the first five years of life. He spoke with a very good vocabulary. Autistic children often have difficulty speaking.

Did he have Tourette's and/or obsessive-compulsive disorder? His motor mannerisms could have been tics. Could the fact that changes in his routine bothered him and induced tantrums have derived from his obsessive devotion to a routine that had been disrupted? This was possible, but the family history was more consistent with a bipolar mood disorder, and indeed his behavior suggested mania.[3]

He seemed to have an internal motor that was racing. His constant moving, difficulty focusing attention, high-spirited inappropriateness, inability to see himself as others saw him, and the lack of caring about the effect he had on others were consistent with mania. His paranoia, manifested by his seeing his sister as a major competitor, was nonspecific. Any of the conditions in the differential could have caused that degree of paranoia, but mania certainly is one of these.

Mania often causes a prolonged sleep disturbance. I asked Mrs. Bloehm if Stanley ever had a problem going to sleep or staying asleep. Her eyes rolled upward. "Yes, indeed," she answered. She told me that he had not been able to get to sleep before 12:00 to 1:00 A.M. for years. She said that she and her husband had had to sit with him to keep him in bed when he was younger. Stanley would often wander around the house throughout the night. After such a night, he could not be easily awakened in the morning when it was time for him to attend school. Mr. and Mrs. Bloehm considered that chronic sleep deprivation might be one of the causes of his poor functioning at school. A prolonged sleep problem of this sort is rarely "reactive" and almost always signals a propensity to serious mental illness.

When Dr. Stein had examined him, Stanley had described odd tastes, smells, and peculiar dreamlike episodes that Dr. Stein speculated could be a kind of epileptic seizure. Could Stanley's whole symptom complex be the result of a kind of complex partial seizure disorder? Indeed, this possibility was the reason Dr. Stein had requested my consultation.

Dr. Stein had prescribed carbamazepine for Stanley. This is an effective antiepileptic medication. Along with other antiepileptic drugs and lithium, carbamazepine is also very effective for controlling mood swings. I asked if carbamazepine had seemed to help Stanley. Mrs. Bloehm said, "Oh, yes! It is the first medication that has ever seemed to help." She said that he was less irritable and showed more self-control during his rages. The outbursts were less frequent and less severe, though they had continued. I suggested that she double the dose of carbamazepine that Stanley was taking. I thought that his symptoms, family history, and response to medication better supported a diagnosis of mania than complex partial epilepsy. This situation is not uncommon. Dr. Stein, a psychiatrist, had made a neurologic diagnosis, and I had made a psychiatric one, but we both ended up using the same medication therapeutically.

Stanley continued to take carbamazepine. His volatile temper was curbed, and he was able to attend a regular class in junior high school. He did not like his sister much and usually ignored her, but he no longer attacked her and occasionally was generous and acted sweetly toward her. He remained a little behind in math though his reading skills rose to the normal range and stopped posing a problem. I followed him for two years.

I considered this to be a real success story and felt some sense of satisfaction for my small role in it, but it is possible that his behavior improved as his brain developed, that carbamazepine had only seemed serendipitiously to work. Perhaps Dr. Stein and I did not help, though the Bloehms credited us with Stanley's turnaround. Even though other physicians, the parents, and school officials thought Stanley had ADHD, I think that his symptoms of ADHD were caused by mania and that the successful treatment of mania made those symptoms resolve.

The Bloehms were extraordinarily good parents and extremely patient with their son. What if they had beaten Stanley every time he manifested bad behavior? There are parents who would have done this to "teach him a lesson that he will never forget." I think that daily corporal punishment could easily have pushed Stanley into a very destructive pattern of behavior that could have ended with his becoming seriously violent.

Children have a limited repertoire of behaviors with which they can convey what is going on inside them. As such, childhood manifestations of bipolar mood disorder are frequently mistaken for other behavior problems. The most common signs and symptoms of childhood onset bipolar disorder are irritability, crying spells, psychomotor agitation, and aggressive temper outbursts. Academic performance may be erratic. What is more, the course of early-onset bipolar mood disorder tends to be continuous and chronic rather than episodic, which is what is normally observed in the adult form of the disorder.[4] Consequently, such children are usually considered simply hyperactive, difficult, moody, oppositional, or just bad. In Stanley's case, the fact that he was being raised in an extraordinarily well-organized, nonabusive household clarified the picture. Teachers and mental health professionals tend to attribute maladaptive behaviors primarily to the home situation.

The behavior of manic, depressed, abused, schizophrenic, and obsessive children can be so similar to that of children with frontal lobe damage or delayed frontal lobe development that the child's behavior does not always differentiate these conditions. To complicate the issue, schizophrenia, obsessive-compulsive disorder, mania, and depression have all been related to a dysfunctional state of the frontal lobes. As with Stanley, the family history of mental illness can furnish an important diagnostic clue as most mental illnesses are genetically transmissable. Suicide, alcoholism, drug abuse, gambling, and wild promiscuity in near relatives suggest a mood disorder. Years of hospitalization of relatives point to schizophrenia. ADHD itself can be familial.

Is Ritalin, the stimulant often prescribed for the symptoms of ADHD, overused? Can it help normal children whose brain development is immature to increase their attention span? Its beneficial effect is similar to that of coffee that helps many adults, who are otherwise normal, function better in the morning and throughout the day. Ritalin might also help frontally damaged children and could even help some of those with depression to increase their attention span. Ritalin would not help with mania, schizophrenia, or obsessive-compulsive disorder and might actually worsen these symptoms. Because the proper choice of medication to modify the symptoms of ADHD is not always obvious, trial and error with antidepressants, mood stabilizers, and antipsychotics may be necessary.[5]

Obviously, most children with ADHD do not grow up to become violent criminals. The very violent individuals I have seen have usually suffered from more than ADHD as children. They also suffered from conduct disorder (CD), frequent manifestations of especially destructive behavior with physical harm to people and property, arson, and cruelty to animals. Oppositional defiant disorder (ODD), characterized by arguments, insubordination, and damaged social relations is also associated with CD. Bedwetting (enuresis) is often associated with CD and often lasts until the teens. CD and ODD are red flags for abuse, both physical and/or sexual as the following story illustrates. It concerns a man I examined for his defense team to see whether any mitigating factors would allow him to be sentenced to life imprisonment rather than execution for having killed one man during an escape from prison and having killed another in a subsequent robbery.

Probably from Ms. Fusco's point of view, Lee was a cute-looking, sandy-haired, freckled eight-year-old in her second-grade class. The photograph I found of Lee in his file at that age was certainly beguiling. I doubt that he was fun for Ms. Fusco in the classroom though. According to his school records, he was going nowhere academically and was grouped with the slowest learners. That was not so bad. He was not likely to have been the only slow learner in her class. What distinguished him was his defiant, angry behavior. Lee had a terrible temper and lost it frequently. He got into fights daily with other children. These fights resulted from arguments over trivia that led to pushing and then fighting.

Lee used physical force, which was unusual at his age. He punched other children in the face. Usually, Lee left his classmates on the ground, crying. He was physically small, but a bully all the same. He had no friends among the children and, according to her report, required Ms. Fusco's constant supervisory attention.

He had broken windows in the school by throwing rocks through them after dismissal. The only break in the flow of this bad behavior came from his frequent, unexcused absences from school, for which Ms. Fusco was probably grateful. She responsibly tried to contact his mother, Marla Dagon, and Lee's stepfather, Tom, with whom Lee lived, but Ms. Fusco could never seem to reach Marla by phone, according to a report she had prepared for the principal.

Marla finally responded to one of several of Ms. Fusco's letters, asking her to come to school for a conference. Marla was intoxicated when she came to see Ms. Fusco, which led the teacher to request a social services investigation.

Lee was found to be living in a filthy, vermin-infested home from which both Marla and Tom were absent most of the time. Marla, her husband Tom, and Lee's biological father, Keith, were all alcoholics whose main social interactions involved drugs, sex, and violence—often directed at one another. Lee had begun sniffing glue and huffing gas at home when he was five. He did this in the presence of his parents! He and his younger brother, Richie, were also forced at a young age (Lee was five; Richie four) to perform fellatio on an older half-brother, Odel, who was their babysitter. I discovered this in an interview with Richie performed by a legal investigator for Lee's defense. Lee confirmed it. Odel indicated that "it might have happened . . . once."

Lee was removed from his home by the state because of neglect, not abuse. Though objective documentation of this neglect dated from his earliest school experiences, it is unlikely that neglect and abuse had started when he went to school. His parents' maltreatment must have started at birth. The state actually knew nothing about the abuse. The authorities placed him in a foster home in a different neighborhood, and he was moved to a new school.

Ms. Fusco's problem was solved by his removal from her classroom. But Lee's problem was far from resolved. In his new foster home, he was beaten for misbehavior and he ran away. He was placed in another foster home, was beaten again, and ran away again. Living in the street, he was arrested for shoplifting. He was released from detention into Marla's care. He voluntarily transferred to Keith's home at times but generally preferred Marla's home as the physical abuse was less severe there. In neither home were there mealtimes, birthday parties, family occasions, or meaningful parental guidance. His parents were absent—working, carousing, living riotously, and selling and using drugs. The only parental tool in their repertoire was beating. There was no tenderness, love, consideration, respect, or decency.

As a nine-year-old, he was supervised by an older half-sister, who acted as his babysitter. She and her friends thought it would be fun to force him to undress and to simulate sexual relations with one of the girls for their amusement. He was told to urinate in her vagina "so that he could become a father." He told his mother, Marla, of this incident and she beat him, he said mercilessly, without otherwise effecting a change in his circumstances.

If she was at home, his mother would beat him daily with a belt she carried around her neck for this purpose. He was totally controlled at home by the whims of older, stronger figures who tortured him and to whom his feelings meant nothing. His father humiliated him by pulling down his pants and beating him in front of visitors. Once he laughed at dinner and some mashed potatoes came out of his mouth. For this, his father beat him with a broom handle that caused welts over his entire body.

He ran away from home and lived in the streets for weeks at a time. He lived by stealing food from supermarkets and prostituting himself to "chicken hawks," men who swoop down on little, homeless boys for sex.

He was arrested again for stealing and placed in foster homes again. One foster placement, the only one he remembered fondly, was with an openly pedophilic homosexual. A social worker who visited the home noted in her report the presence of men wearing makeup and dressed as women. Because this foster "father" did not physically force him to perform sexual acts and provided enough food to prevent hunger, Lee voluntarily returned to this home several times over the years, although never permanently.

Often arrested for shoplifting and theft, Lee told me that he was returned to Marla or Keith's home by the police. There he would be beaten severely. He was again taken to shelters, foster homes, and detention centers where beating and forced sexual acts were the rule. He ran away again. When he became angry, and he got angry easily, he hit others, hurt them, and destroyed property. He told me that he engaged in arson by setting fields on fire. He had enuresis until he was eleven years old. Until age sixteen, when he dropped out, Lee's school performance was characterized by truancy, fighting, mischief, destruction of property, and inattention to his studies—in other words, ADHD with conduct disorder.

ADHD with CD is probably the matrix from which later serious violence springs. It is a childhood behavior syndrome with essentially the same determinants as adult violence and is a prelude to violence. It should sound alarm for teachers, social workers, physicians, and school nurses that a social investigation is urgently indicated.

Of what should a social investigation consist? The primary focus must be a search for abuse. It is amazing that many clinics devoted to the care of schoolchildren with behavior problems do not address abuse. Lee was removed from his mother's home and still there was no comment in the social worker's reports and records concerning the physical and sexual abuse about which he and his brother told me later on. Officials and professionals who work for the state usually skirt the issue, talk around it, and somehow never get to it. As demonstrated by Lee and his mother, Marla, the abusing parents and the abused children both have some motivation for camouflaging it. For different reasons, both parents and children expend energy to keep abuse from view. Both may also have some desire to obtain help, and, with sensitive questioning and investigation, the full story can be brought to light.

We do not know how to determine the best way to reverse or prevent abuse, and how to estimate the effectiveness of intervention. It seems likely that the earlier childhood abuse is discovered and ended the better. It is not known if there is a time when intervention is too late to protect the child from the risk of becoming a violent adult. Whatever the time may have been for effective intervention, that time had long passed for Lee. Permanent incarceration and execution were the only options that remained.

CHAPTER 7

Toxins and Turmoil

Not all murderers are poor like Dirk Donovan or African American like Cynthia Williams and Bobby Moore. The same factors that interact to cause violence operate malignantly at every social level and in every racial group. Abuse, mental illness, and neurologic deficits, not poverty or race, cause violence. Toxic encephalopathy produces neurologic deficits and the symptoms of mental illness.

Ray Hawkins, an eighteen-year-old busboy, rode his ten-speed bike to a convenience store at midnight. He was drunk and wanted to get some more beer and some cigarettes. At the restaurant where he worked, he had already consumed a fifth of bourbon and many beers during his eight-hour work shift, but he needed more alcohol to sleep. The female clerk was not behind the counter when he entered the store. He walked behind the counter and was helping himself to a carton of cigarettes when the clerk returned and approached him from behind. She probably tapped him on the back and asked what he was doing behind the counter. He wheeled around and savagely stabbed her repeatedly with an enormous hunting knife that he always carried. Some of her wounds were defensive. Many were potentially fatal. Her heart, lungs, aorta, and neck were pierced. She collapsed, dead, behind the counter in a pool of blood. He walked though her blood to the refrigerator in which beer was kept and then to the door and off into the street, leaving behind him a trail of bloody footprints.

A subsequent customer found her body and alerted the police. Amid the hubbub of police cars, radios, flashing lights, ambulances, and bystanders that followed, Ray, riding his bicycle, unsteadily returned to the convenience store. He and the police shared a moment of mutual recognition. He turned and quickly pedaled away from the pursuing police, who recognized Ray as a juvenile delinquent who often pilfered beer and cigarettes from convenience stores. Outdistancing the police by crossing a pasture, Hawkins abandoned his bicycle and, in the darkness, ran to his parents' home through some woods. Without making any attempt to hide his bloody clothes or knife, he dropped the knife in the garage, entered his adjacent room, and collapsed on his bed in besotted but deep sleep.

Early the next morning, several police officers came to his parents' home and awakened him. His bloody clothes spoke for themselves, and the blood-covered knife was lying openly on the garage floor. On being shown the knife, he fainted. He was arrested, tried, convicted of first-degree (felony) murder, and sentenced to death. The sentence was appealed on grounds that his counsel had been ineffective. No mitigating circumstances had been presented or sought by the defense.

The new defense attorney arranged for me to examine the boy as a second sentencing trial had been ordered. The attorney asked me to determine if any mitigating facts existed. A thin, pale nineteen-year-old man with strawberry-blond hair was led in shackles into the interview room. After unshackling Ray, the guard withdrew and observed us through a window. The boy's freckles, large blue eyes, and friendly, enthusiastic manner projected a youthful innocence that was jarringly contradicted by a homemade tattoo on his forehead. He was extremely polite and constantly used the word "sir" to address me, according to southern custom. He had recently experienced a religious change and felt "saved." He made references to the Bible and God in almost every sentence and, as he warmed to me, he became so enthusiastic about what he was saying that he literally jumped up and down in his chair.

This caused the guard to increase the intensity of his observation, but the boy produced no sense of threat. His body language was unusual, and what is unusual is often abnormal. It suggested immaturity and a defect in his ability to modulate impulses.

With the exception of an occasional double negative and mistake in number ("that don't bother me none") his vocabulary was good, his conversation well directed, and his intelligence normal. He certainly was aware that he was under sentence of death and that I had been engaged by his defense lawyer who was attempting to save him from the electric chair by presenting mitigating factors. The fact that Ray had repeated kindergarten and been slow in the subsequent development of reading and writing skills was consistent with the theory that he may have suffered some early neurological dysfunction but meant little on its own.

I started with the family history. I knew that his father, Jack Hawkins, had died and I asked, gently, about his father's illness and what sort of a man his father had been. His demeanor changed as we talked about his father. He became serious and almost seemed frightened. He said, "My father was a good man." He seemed to be trying to convince himself as well as me as he repeated, "He was a *good* man."

I expected to find a history of physical abuse dating back to early childhood, but no one offered information to me confirming that conclusion. Far from it, Ray said he had not been abused. In fact, Ray's account, consistent with the story his mother, brothers, and stepfather told his lawyers, was romantic and melodramatic. According to this legend his father, Jack, had died tragically of leukemia. The untimely death of such a "good man" and "good father" had unbalanced Ray. Of course, after he had returned from heroic duty in Vietnam, Jack was, understandably, a "somewhat changed" man and as a result might have occasionally been "too strict" with the children. But he was good! So good!

Then Jack developed leukemia. Leukemia caused him to drink heavily but despite his drinking, he remained a good man, a good husband, and a good father who was especially good to Ray. Jack had taken trips alone with Ray when Ray was a little boy and Ray grew and developed basking in the warm sunlight of their excellent relationship. The death of his father had caused poor Ray to lose his moorings. He imitated his dad and began to drink. The murder he had committed was the effect of alcohol on a good boy who had become mentally unhinged by his father's piteous death. While he recounted this version of the causation of his crime, the boy's facial expression suggested that his account was not true. His eyes wandered as if he were seeking inspiration from above.

The investigator and the entire family specifically denied any abuse by Jack and refused to indicate what they meant by his having been "too strict." The party line was clear. No one was really to blame. Ray, his mother, his brothers, and his stepfather were actually reconciled to Ray's execution in order to preserve their myth of his "*good*" dad, Jack Hawkins.

I met with Hawkins for three hours and then separately with his mother, brother, and stepfather for another two hours. By repeated questioning and following conversational leads beginning with the phrases "too strict" and "changed man," I was able to expand my understanding of the situation.

Ray was one of four children born to an extremely dysfunctional, white, middle-class family. Ray's mother, Alice, was Jack Hawkins's second wife. Jack's first marriage had been marked by many brawls. Jack thought that his first wife was poisoning his food and was seeing other men. This paranoia led to fights and, after about three years of wife-beatings and calls to the domestic relations unit of the police, his first wife divorced him. Jack remarried soon after and Alice bore his first child, a daughter. When this child was a baby, Jack divorced Alice because he thought that she, too, had been "running around" and poisoning his food.

At the time of this divorce, Alice allowed Jack to take her baby daughter away from her. Jack gave his daughter to his brother, who subsequently gave her to Jack's mother, who lived in another state. After a brief period in her home, Jack's mother gave the little girl to the child's paternal aunt. The girl was raised by her aunt in another part of the country about one thousand miles away. Alice had been given permission to see her own daughter only once or twice since that time, despite her desire to visit the child more frequently. This seemed to exemplify Alice's ineffectiveness in family matters. While relating the story of the abduction of her daughter by her husband, Alice was wracked with heartbroken sobs.

Alice and Jack got together again shortly after Jack removed their infant girl from their home, but they never formally remarried. Ray and his two younger brothers, Arnold and Darren, were born after that "reconciliation."

By the time Ray was eight years old, Jack had degenerated into an abusive alcoholic who daily beat his wife and children. As his consumption of alcohol increased, so did the severity and frequency of the beatings.

Alice's cooking and cleaning were "never good enough" for Jack. Characteristically, Jack and the boys ate in the living room in front of the television set. Alice did not join them for fear that she would be struck. Typically, Jack would say, "This food stinks!" and he would take a plateful of food and throw it at Alice or against the wall. He would then brood opposite the TV, drinking beer while the children shivered in fear.

As Ray recounted this to me, he wept profusely. When I told Alice what Ray had told me, at first she denied it, but then tears rolled down her cheeks and she sobbed for a long time. She admitted that it was true and then she added further details.

On one occasion, a terrible fight developed between Alice and Jack in which Jack beat Alice's head against the floor. She called out in mortal fear to Ray, then about eight years old, pleading with him to get a knife and help her by stabbing his father. Terrified, Ray ran into his room, and closed the door, not complying with his mother's urgent request. A neighbor heard the screaming and called the police, whose intervention saved Alice's life. This kind of interaction was not isolated but repetitive over the course of years. "Get the knife!" was a maternal injunction that resonated in Ray's mind over the years.

His father beat Ray with sticks, belts, buckles, a guitar, and a gun barrel. He also took Ray on long car trips away from his home, which were initially, presented to me by the family as vacations, an example of what a good relationship Jack had with his son, Ray. In fact, these were separations from Alice, initiated by Jack in an angry, intoxicated, and abusive state. Jack wanted to punish Alice by abducting the terrorized Ray from his own home, removing him from his mother's care, and abandoning him to the care of Jack's mother in another state, never to be seen again—what Jack had done with Alice's oldest daughter. Sometimes these trips included Alice, whom Jack wished to degrade by forcing her to give up her child, with her own hands, to her mother-in-law's care.

On one such trip, Ray remembered sitting in the back seat with his mother while his father drove. He was nine years old, and he was crying about the prospect of being forever removed from his mother. Jack became enraged, saying, "I'll give you something to cry about!" He stopped the car and beat Ray with a whip that was kept in the car just for such occasions. The beating was delivered to his bare back, buttocks, and legs, so viciously that the blows drew blood. After the thrashing, his mother tried to comfort him and treated his wounds by pouring rubbing alcohol on them.

When I asked, "Didn't that hurt?" he responded, turning pale at the recollection. He rolled his eyes to the ceiling and said, "Lord have mercy!" There were many confirmatory horizontal scars on the boy's back and thighs that derived from savage whippings that had been delivered over the years.

Shortly before leukemia claimed him, in a desperate attempt to prolong his life, Jack traveled to the Bahamas to receive Laetrile, an ineffective nostrum for cancer that had been banned in the United States. Alice and Ray went with him. After Jack had received this "treatment" for several days, his condition seemed to stabilize. Alice, tired of being cooped up in the small apartment they had rented, indicated that she wanted to go to the beach with Ray. At this, Jack "pitched a fit." He wanted Alice to stay in the hotel with him. Alice defied him and told him "to like or lump it" and went to the beach with Ray.

When they returned, there was a fistfight between Alice and Jack. The twelve-year-old Ray stood up for his mother and said to his father, "Why don't you die?" Within a week, Jack was dead, and Ray felt extremely guilty, as though he were responsible for his father's death.

The stress that Ray was under at this time is evidenced by his numerous visits to the army base medical dispensary, which served the family as a primary care center, for minor medical complaints like headaches, stomachaches, and constipation. During the year prior to his father's death, Ray's visits to the dispensary increased from once a month to several times a week. After a few months, no notations were even made in Ray's chart about the reason for the visit beyond the fact that he had been to the dispensary. No one in that medical facility asked about or discovered the cause of Ray's apparent stress. There was no referral to a mental health clinic. Ray's abuse, Jack's drinking, and the spousal abuse were all shrouded from the view of the medical personnel. Needless to say, no attempt was mounted to intervene, though some of the doctors' cryptic notes indicated that they were all well aware that Ray's problems were solely emotionally based.

As Jack became sicker and sicker with leukemia, he became more and more abusive and he drank increasingly. Ray dated the onset of his own drinking problem to the time of his father's death when he was almost thirteen years old. Ray patterned his drinking after his father, consuming a twelve-pack of beer and a pint of whiskey over a forty-five-minute period. This would raise his blood alcohol concentration to an estimated 0.3 percent, three times the legal limit of 0.1 percent in most states. Ray then continued to drink steadily all day, every day, from the time he got up in the morning until the time he went to bed at night. He often held jobs despite this, but his life became progressively more disorganized as he succumbed to the effects of alcoholism over the years.

He began to have "blackouts," periods that he did not recall while drinking, but during which he seemed to others to be behaving normally. As his addiction advanced, Ray developed a tolerance to alcohol. Even his mother had difficulty telling if he were drunk. She said she could usually detect it by his bloodshot eyes and the occasional slurred word. A stranger might have thought he was sober, as he could walk a straight line even while drinking heavily. Yet his hands shook every morning so badly that he could not write. Morning tremor (the shakes), a sign of alcohol addiction, is caused by several hours of abstention from drinking and indicates the beginning of the withdrawal state.

Ray had a longstanding paranoid tendency that was marked even before he began to drink. He had never developed any meaningful, long-term friendships. He said that people outside the family tended to avoid him. His father told him that if anyone at school threatened his younger brothers, Ray was expected to fight to protect them. There were many fights. He was a bully. The result of this was "everyone hating him."

He felt blamed for things that he did not do and that people were out to get him. He always sat with his back to the walls so that he would not be attacked from behind. This was so prominent a personality feature that long before the crime and before he began to drink, even while he was in primary school, children used to tease him by coming up behind him and tapping him on the shoulders. He would gratify them by leaping into the air in terror. He frequently imagined being threatened and cowered at home, saying to his mother, "They're after me, if I go outside they'll get me."

He often seemed scared without having a focus for his fears. He slept at home with knives around him and next to him and frequently carried a knife with him when he went outside, for protection against unnamed threats.

The combination of alcohol and paranoia was unpredictable. When he was alone, alcohol calmed him. In social situations, alcohol increased his irritability and caused tremendous lability of mood, with shifts from happy to sad, and angry to violent. At times he even became suicidal while under the influence of alcohol and put himself in danger by jumping fire pits on a bicycle or by crossing busy streets on foot without looking for oncoming automobiles. Depressed individuals who are ambivalent about suicide often "leave their fate in God's hands" by taking risks—

driving dangerously, playing Russian roulette, and precipitating fights.

Ray was first arrested for driving while intoxicated at age sixteen. He had wrecked cars on four occasions between the ages of sixteen and eighteen while under the influence of alcohol. His mother repeatedly enabled him to obtain a new car until he lost his driving license because of repeated DUI arrests. At that point, he had to use a bicycle to travel to work.

At age sixteen, after he created a dramatic scene with wild, drunken behavior at a neighbor's house, he was taken to an alcohol treatment center. At the time of his admission, he weighed 110 pounds but should have weighed 142 pounds considering his height and age. I calculated that he needed 1,650 calories a day to maintain his ideal weight. Because he was getting more than 1,400 calories a day from beer, only trivial amounts of his total caloric intake came from nutritious sources. Most of the time, he was essentially starving, only obtaining calories from beer. He ate food in binges, going several days at a time without eating.

Long periods of addiction to alcohol cause degeneration of parts of the brain, especially the portions with high metabolic rates. The cerebellum and the cerebral cortex are especially likely to be damaged by heavy drinking. Permanent damage results from both nutritional deprivation (starvation) and the direct effect of alcohol on nerve cells. A hospital examination revealed that Ray had liver spots (telangiectasias) over his shoulders, chest, and upper back. His liver was enlarged with cirrhosis. Blood tests that reflect liver function were abnormal. He had pancreatitis and was vitamin deficient.

He was placed on a proper diet with vitamins, sedatives, and tranquilizers. The sedatives and tranquilizing drugs were gradually withdrawn, and he participated in group meetings. After two months in the alcohol treatment center, Ray was discharged, "cured."

Within three to four months after his discharge from the treatment center he began to drink again. His hands again shook every morning and sometimes he spoke of feeling that bugs were crawling all over him. He hid bottles of whiskey and beer all over the house, in his bed, behind the refrigerator, under his mother's bed, and he experienced frequent blank spells.

On the very day of his discharge from the alcohol treatment center, coincidentally his seventeenth birthday, he was given a gleaming and inappropriate prize, a giant hunting knife. This was to become the murder weapon. The knife was a gift from his brother, Arnold, with the permission of his mother. "He loved it," his mother told me. He carried it everywhere, hiding it in his side boots, ankle, jacket pocket, or in his back belt, "for protection."

It did occur to Alice and her new husband Lew that it might not be a very good idea for Ray to have this large, lethal weapon or for him to carry it around with him all the time, but neither intervened. Ray continued to carry it, and they did not remove it from their home because he "liked it so much." On several occasions, afraid that Ray might use the knife while in an alcohol-induced rage, Alice and Lew actually confiscated it and hid it in their house, but Ray searched for it and ultimately found it on each occasion. In fact, there was a near miss that foreshadowed the murder.

Several months prior to the convenience store homicide, Ray was at home alone with his brother, Arnold. Arnold thought that Ray had consumed too much liquor that day and evening and hid Ray's remaining bottles. At about midnight, Ray pushed Arnold in anger because he wanted the liquor to be returned. When Arnold pushed him back, Ray left the room, went into his bedroom, took out the giant hunting knife and "came at" Arnold with it, his face contorted in a frightening manner. Arnold quickly placed a chair between them and held his hands up in the air as if in surrender and said, "Go ahead and kill me." Ray hesitated and lowered his head and eyes toward the floor, as if ashamed. At that moment Arnold picked up the chair and broke it over Ray's head. Ray dropped the knife. Arnold quickly retrieved the knife, ran with it outside the house with it, and threw it into the family's large backyard swimming pool. The two brothers then wrestled, fighting quite seriously for fifteen to twenty minutes until Arnold was able to escape from Ray.

When he returned home about four hours later, he found Ray passed out on the floor of the hallway. Arnold left him there undisturbed, all night. Arnold locked himself in his own room and went to bed. The next morning, all was calm. Ray asked Arnold what had happened and why he had apparently spent the night sleeping on the floor in the hallway. He had no recollection of the events of the prior evening and Arnold did not tell him, nor did Arnold tell his parents when they returned from an out of state trip.

My neurological examination demonstrated that Ray suffered from brain dysfunction. He could not focus his attention. There was motor impersistance and discontinuous jerky movements of his eyes during visual pursuit. He was clumsy (apratic) in performing patterned repetitive motor movements. The unsteadiness of his stance and gait also indicated damage to the cerebellum (the part of the brain that controls coordination and is often damaged by chronic alcohol consumption). There were choreiform (discontinuous, jerky) movements of his outstretched hands that derive from damage to subcortical centers. Thus there was evidence of brain damage with frontal lobe signs, cerebellar and subcortical abnormalities. The acute effect of alcohol consumption would have been to further reduce his capacity to control his impulses. He was more sensitive to alcohol than he would have been had he not been permanently brain damaged.

This young man had developed the physical stigmata of alcoholism by the time he was sixteen. The phenomena of tolerance, withdrawal symptoms, and a psychological need for alcohol confirmed the diagnosis of addiction.

His erratic activities on the evening of the murder, his only partial recollection of the events, and his return to the scene of the crime on bicycle to determine what had really happened were consistent with the kind of wild, confused behavior that he had sometimes manifested while intoxicated.

Even sober eighteen-year-olds can have difficulty controlling their impulses because their brains have not fully developed. Their frontal lobes may not be fully hardwired. It may be strictly incorrect to equate immaturity with brain damage, but the fact is that normal, adolescent behavior is a great deal like that of adults who have sustained frontal damage. The frontal lobes normally are not fully myelinated until the early twenties. Immaturity literally means undeveloped. But Ray's brain had also been damaged by alcohol. Neither the impulse to kill nor his incapacity to resist it were under his full control at the time.

He had just turned eighteen when he committed the murder, in a wild, paranoid state, his mind and his capacity to inhibit his impulses and memory dulled by alcohol. During his alcohol-induced encephalopathy, his murderous, paranoid response could not be checked. He had been abused and his behavior was deeply influenced by paranoid thoughts. The judge agreed that these factors were mitigating and commuted the sentence of this nineteen-year-old boy to life imprisonment without the possibility of parole.

Ray's family was white and middle class. They owned their home, had a swimming pool, and had cars, clothing, and food in abundance. The children received presents and lived with both parents. Yet, "something" was wrong; the same "something" as in the other cases I have discussed: abuse, paranoia, and brain damage. The alcohol that influenced Ray at the time of the murder reduced his capacity to control his impulses. Drugs, chiefly alcohol, are factors in up to 70 percent of homicides (cf. reference 7 in chapter 3).

The acts of all the murderers I have described so far were different. A girl killed a fellow student in a dispute: a man overkilled two victims in the course of planned robberies; another assassinated three victims in a failed robbery attempt; another participated in an opportunistic rape/murder and impulsively killed a police officer during an unsuccessful robbery; and Ray killed a clerk in an alcoholic haze in pursuit of alcohol and cigarettes. Two of these murderers were white; three were from the lowest socioeconomic groups; and two, one white, one African American, were middle class. In each case, abuse was extreme, but in three of the four cases, extraordinary efforts were necessary to uncover the history of abuse. In our first case (Cynthia Williams), we did not obtain a full description of abuse because we had no idea how to do this nor did we fully understand the importance of abuse as a causal factor in violent crime. All five individuals were paranoid and all were neurologically abnormal. Thus, the similarities in these seemingly disparate cases are, in fact, more striking than the differences.

CHAPTER 8

Nature, Nurture, and Neurology

We are all interested in the concept of biological determinism, the idea that the brain directs behavior and that genes and experience imprint the brain and determine what we do and what we think. When we jest and say that men do not ask for travel directions, react angrily to challenge, or refuse to participate in household chores, we jokingly attribute these characteristics to the Y chromosome. But every joke has a grain of truth: the male ego, male aggressiveness, male sexual drives, and the male need to dominate are the subjects of *bon mots* and quips, but at some level we are willing to accept the proposition that the origin of some of the differences between men and women is biological.

Just as the average person is willing to attribute gender differences to biology, we have become increasingly willing to accept the proposition that personality features and intelligence are inherited. This represents a change in public attitude. About thirty years ago, a huge controversy erupted over the suggestion that intelligence was heritable. Now no one blinks at the suggestion, reported in *Newsweek*, that genetics account for 50 percent of the differences among IQ and 30 percent of the differences among personality.[1] Anything that is genetic originates in the billions of coded DNA sequences we have inherited from our parents. In a study known as the Human Genome Project, scientists at the National Institutes of Health in conjunction with private companies are racing to decode all human genes, to identify each gene and its role in the body.

Many reports have been published that claim to have found the gene responsible for certain common psychiatric disorders like schizophrenia, mania, depression, autism, and obsessive-compulsive disorder. Though there have been no universally convincing successes and controversy continues, both failed and successful attempts to provide molecular genetic correlations with clinical behavioral conditions seem to have strengthened the conviction that an unequivocal answer will be found, and that a genetic defect is responsible for each disease in question.

Many of the behavioral diseases that are now considered to have a genetic basis were just a few years ago treated as purely psychiatric disorders of the mind, rather than the brain. The distinction between the mind and the brain that so recently seemed so precise to both doctors and laypeople is disappearing.

Odd behavioral fragments are associated with genetic and chromosomal disorders, implying that abnormal behavior and the urges that drive it are also encoded in the brain by genes. For example, Tourette's syndrome is a hereditary disease characterized by tics, vocalizations, and obsessive-compulsive symptoms. In the Lesch-Nyhan syndrome, a genetic disease caused by an absent enzyme, sufferers have the uncontrollable urge to mutilate themselves. The Prader-Willi syndrome, caused by the deletion of genetic material from a chromosome, involves uncontrolled appetite, gorging, obesity, and constant thoughts about food. It is easy to accept the proposition that obsessive-compulsive symptoms, self-mutilation, and indiscriminate, voracious appetite evident in these disorders are propelled by a poorly constructed brain. It is only a short step beyond this to conceive of all obsessive-compulsive symptoms, all self-mutilation, and all uncontrolled appetite and obesity as the result of a subtly disordered brain anatomy and physiology, even when these symptoms are isolated and not part of one of these genetic conditions.

However, when the individual's obsessive thoughts involve violence, when the mutilations he practices are directed toward others, when his insatiable appetites involve criminal sexual practices, we are repelled and quail at the proposition that these abominable thoughts, urges, and acts originate in the brain. When immorality, especially violent serial murder, is the issue, we are much less likely to accept the brain's biology as the cause. We do not want to allow offenders to escape punishment, and we want to maintain the line that separates victims and perpetrators.

Instead, we invoke the concept of free will as the factor that controls the actions of criminals. When people speak of the "mind," "psychological factors," and "volition," they usually mean something outside or even above the brain. We do not want to accept biology as the only determinant of morality, but it is surely one of the determinants.

The biology of the brain is not shaped only by genetic influences. What the brain registers through its sensory systems about the surrounding environment is increasingly recognized as a critical factor that permanently changes the brain by altering its connections.

There are two great oak trees in my front yard, each about two hundred years old. One stands straight and tall in the front yard bathed in sunlight all day, just south of my house. Its branches and leaves are distributed symmetrically in a large, majestic circle. The other is located at the lip of a declivity that leads down to a lily pond about ten feet below the level of the front yard on the east side of the house. The second tree bends at a precarious angle in the direction of the pond and its primary branches curve asymmetrically backward and upward toward the sun. Its afternoon sun is blocked by my house and its main leafy skirt stretches toward the warmth and light of the morning sun.

Both trees are magnificent, but each has a different configuration of its trunk, branches, leaf distribution, and, probably, root system. Each grew from a little acorn that genetically was very much the same as the other, if not identical. The differences in shape between the trees reflect the slightly different environment they encountered on my half-acre of land. Here is the effect of environment on the expression of a genetic endowment. By means of an arboreal sensory system, each tree had detected the environment and adapted to the presence of water, sun, and nutrients in the soil.

How like a tree a nerve cell seems. Each nerve cell sends out many branches and ultimately develops about ten thousand connections (synapses) with other nerve cells. Some of these synaptic connections receive stimulatory or inhibitory influences from other cells (dendrites). Some of its synaptic connections send information to other nerve cells (axons). Geometrically, the nerve cell is like a tree and the root system and leafy branches are like the branching dendrites and axons. The nerve cell extends upward and downward in three dimensions. The outside environment registers its influences on the development of the young cells, which, like saplings, start with very few branches.

The development of branching is the trees' response to their genetic instructions and to what they sense in the environment. The development of synapses between nerve cells is quite similar—the result of genetic instructions and sensory input during early life.

The sensory systems of the brain detect sights, sounds, odors, tastes, pain, touch, temperature, pressure, motion, and position. The development of the synapses between brain nerve cells is very sensitive to the outside environment, including the psychosensory environment.

I like this metaphor, but metaphors do not prove anything; they are just illustrations. There is also a familiar clinical condition in which a genetic endowment had been changed or lost as the result of sensory deprivation. If an infant is born with a cataract, it must be removed early or vision will not develop in that eye. Visual impairment can affect the sensory input and permanently change the organization of the infant's brain. This is not true of adults. If dense cataracts are removed years after they have formed from the sightless eye of an adult, vision can be restored.

The sensitivity of the immature nervous system can be shown with another disorder of the visual system. For instance, a "lazy eye" in childhood will cause permanent blindness if it is not corrected before the child is six years old, even though plenty of light gets into each eye. The medical term for the blindness is *amblyopia ex anopsia,* which means blindness from not seeing. *Ambly-*

opia ex anopsia is a tongue twister but a great example of the point I want to make.

The condition starts with a weak eye muscle in an infant who is born with normal vision in both eyes. Having a weak eye muscle produces double vision. To counter this, the child's brain suppresses the image that comes from the weak eye by means of a cortical (brain) process that is not completely understood. Patching the good eye before five or six years of age forces the child to use the weak eye for vision. Otherwise, the child will become permanently blind in the weak eye.

This form of blindness only occurs in childhood. Adults never develop *amblyopia ex anopsia* even if they develop the equivalent of a "lazy eye" and have double vision as the result of some disease of the eye muscles. The adult brain can suppress the image that comes from the weak eye by means of a cortical process similar to the one that operates in young children. In this way the adult avoids double vision but the eye with the weak muscle never becomes blind.

Something like this is probably the very brain phenomenon that occurs in emotionally deprived children. It was what had happened to Louis Culpepper, to Cynthia Williams, to Bobby Moore, to Dirk Donovan, and to the other criminals described in this book. Their brain development had been skewed permanently by their horrible, abusive upbringing. The impulses and urges initiated by abuse were encoded in their brains. If they were to lead noncriminal lives, their brains would have had to inhibit those impulses and urges.

The vulnerability of the immature nervous system to permanent, irreversible changes as the result of sensory deprivation or manipulation probably underlies the common observations that "children learn quickly" and that "youth is impressionable." The psychosensory environment can cause permanent changes in the brains of children for good and for bad. Genetic endowments, present at birth, can be permanently rendered ineffective or can be maximized depending on the earliest sensory experiences of the child as his or her brain is developing. The issues are not nature versus nurture, the brain versus the mind. The mind is the brain. Both nature and nurture direct its development, anatomically and physiologically, just the way nature and nurture can affect the ability to see.

People with frontal lobe lesions or with damage to parts of the brain that connect to the frontal lobes often cannot prioritize or concentrate very well, symptoms shared by many people with mental illnesses. Frontal lesions caused by trauma, tumor, and strokes are easy for radiologists to see with MRI scans and for pathologists to identify when they study brains at autopsy. Mental illnesses have eluded radiologic and microscopic definition. Schizophrenics, manic-depressives, and those with some very severe neurological disorders like mental retardation, movement disorders, and epilepsy all have normal brains, according to radiologists and pathologists. But obviously their brains cannot be normal. Why have such major diseases of the brain escaped anatomic definition? Because physicians have been looking in the wrong place. The synaptic connections between nerve cells are where the medications work that modify mental illness, movement disorders, and epilepsy. The synapse is the likely site of abnormality in these disorders, but it cannot be visualized by radiological or routine neuropathological techniques.

When we look at trees, we can obtain a qualitative and quantitative estimate of their branching structure. In this way we can estimate the impact of the environment on the growth and development of the tree. If there is a disease of trees in a forest, botanists can examine several trees extensively, noting their branching, their leaves, and their trunks; they can also see the forest as a whole by walking through it and by observing it from the air.

We cannot see the branches of nerve cells in this way. When pathologists look at a little piece of brain under a microscope, the stains they use only allow them to see two-dimensional cell bodies with nothing between them. It is as if a forest were clear-cut and an aerial photograph were taken of a small sample of the stumps. It would be impossible to appreciate the three-dimensional branching structure or to measure the effect that the environment had on the development of the branches of the trees, just by looking at the photograph of the clear-cut forest or a two-dimensional slice of one or two trees. Pathologic examination of the human brain has not provided an explanation for *amblyopia ex anopsia* or for the behavioral effects of a terrible upbringing. Perhaps it never can, with its current tools, but the impact of sensory deprivation on the development of the brain can be studied in animals.

The anatomy, physiology, and function of vision in animals can be permanently altered by temporarily closing an eye or by producing double vision. This phenomenon was shown by two Nobel Prize–winning scientists in studies they conducted on kittens and baby monkeys. Similar changes in the hearing and feeling (somatosensory) systems follow sensory deprivation of animals early in life: permanent defects in these sensory circuits only result from the deprivation of immature animals.[2]

The only way babies or toddlers can learn about the world is through their sensory systems. Since vital functions of the brain can be lost or changed as the result of abnormal sensory input, it is likely that child abuse or neglect at an early age, delivered repeatedly over a period of years, has a major effect on the actual organization and function of the brain. This is difficult to prove, however, because pathologists have limited tools and human experimentation raises ethical concerns.

An enriched and stimulating environment enhances the development of contacts between nerve cells in animals. For instance, use stimulates and nonuse retards the cellular development of the motor system in the brain. Rodents and dogs that were isolated from their mothers and peers early in life demonstrated later behavioral neurochemical, anatomic, and electrophysiological defects. The vulnerability of the nervous system of animals to changes in the environment is especially marked in early development.

There are strong implications in these animal experiments for human brain development. These findings in animals provide an experimental context for the understanding of the effects of the early psychosensory environment on the development of the brain/mind in humans. In the 1940s and 1950s, Rene Spitz, the famous French psychiatrist, Sally Provence at Yale, and others described a delay of motor and social development in institutionalized infants who had been deprived of maternal contact in the first year of life.[3] Dramatic improvement followed when they were provided with the benefit of personal, individual maternal care, but this did not entirely reverse the impairments. These persisted throughout the preschool years. Deficits remained, especially in those children who had been institutionalized during their first two years of life.

Spitz and Provence described abnormalities in the institutionalized children in their capacity to form emotional relationships and to control their impulses. They remained defective in areas of thinking and learning that reflect adaptive capacity and imagination. These cognitive capacities are normally importantly mediated by the frontal lobes of the brain, the last part of the human cortex to mature.

Why is the immature nervous system so susceptible to environmental stimulation or the lack thereof? At birth, human infants are completely helpless. They can breathe and their hearts beat, but they cannot sit, crawl, walk, speak, or reason. About all they can do is eat, sleep, and cry. This is because the brain has not matured yet, even though the number of nerve cells is complete. During the course of childhood, nerve cells in the brain begin to make contacts with other nerve cells, advancing the process of myelination. This refers to the deposition of a fatty insulation material that covers nerve fibers; without it they cannot carry electrical impulses and therefore cannot contact other cells. Myelination of the back of the brain, the occipital lobes, is nearly complete at birth. Gradually, myelination advances forward, with the frontal lobes the last to be fully myelinated. The process of myelination is not complete until about twenty years of age.

The immature nervous system is like a forest of young saplings. The number of trees is about the same in youth and maturity—but what a difference the full development of roots, branches, and boughs makes to the forest as a whole! If the environment is too cold, dry, wet, shaded, or lacking in nutrition, the woods will not develop to the limit of the genetic potential of its trees or their adaptation will be skewed.

Like the trees of the forest, the nerve cells of an infant respond to the sensory stimuli of the environment. Children learn what is happening around them and the development of their nervous systems is dependent on the quality and quantity of sensory stimulation. Scans of severely neglected children have shown that certain parts of their brains are underdeveloped. Studies indicate that if a baby is not held and touched and spoken to and given all the sensory stimuli that are manifestations of parental love, part of the child's brain is not organized properly. The brain wiring of a baby who has been neglected or abused is probably abnormal, less dense, and less complete.[4] Of course, traumatic brain injuries in abused children are common, frequent, and of great consequence.

The behavioral and cognitive effects of early deprivation can be seen in Romanian and other Eastern European orphans, many of whom have been adopted. There appears to be an increasing chance of severe behavioral problems that depends largely on the duration of time the children have spent in orphanages in their earliest years. These behavioral and cognitive abnormalities do not always completely disappear when the child is transferred to a warm and loving environment. There is a window, the first two years of life, during which an infant must receive a parental "love bath" or be at risk, permanently, of suffering the effects of this early deprivation.[5] Similar responses have been recorded in monkeys.[6]

Deprivation can take many forms. One is physical neglect. Emotional neglect is more prevalent both within institutions and at home. Parents and parent substitutes must hold, cuddle, talk to, sing to, rock, play games (like peekaboo), and interact with infants in a loving and supportive manner—not just keep them dry, warm, and well nourished. The absence of these manifestations of parental love is harmful, but even worse is physical and/or sexual abuse. Delivered at an early age, abuse has a destructive reach that extends into later childhood and far beyond.

A famous study revealed the effect of abuse on the behavior of toddlers.[7] Of twenty toddlers in a playschool group, ten had been seriously abused and ten had not. The nonabused children were described as responding constructively when a fellow classmate cried in distress. They showed concern or interest and often tried to provide some comfort. None of the abused children showed concern or tried to help. They manifested distress and fear. Some threatened, manifested anger, or actually struck the crying child. Abused children often display short attention span, overactivity, defiance, fighting, a desire to break things, and they engage in risky, impulsive antisocial behavior. These features persist. Abuse has been implicated in the development of antisocial personality disorder and other behavioral syndromes (borderline disorder, hysterical conversation reactions, pseudoneurological disorders, and dissociative disorders).[8] The corporal punishment of adolescents is a risk factor in the development of depression, suicide, alcohol abuse, child abuse, and wife beating.[9]

So often does violence seem to be transmitted across generations that some have speculated that there is a gene for violence. Personally, I doubt that such a gene will ever be found. If I am right, there must be some alternative explanation for the intergenerational cycling of violence.

Actually, there is an intergenerational aspect to many child-rearing practices, even benign ones. For example, I recently observed my three-year-old grandson playing a little boisterously with his good friend. My son, his father, called out, ineffectively, to the little boy, "Pretend you have already been hurt and stop it before you really have been hurt."

I was gratified, in a way, because I had used those exact words, ineffectively, when my son had played boisterously with his little brother many years before. As a matter of fact, my mother used the same expression when she addressed me and tried to prevent me from tripping my little brother during our play when I was a boy. Not only was her injunction ineffective, but I had hated it when she used those words. She said that her father used to say this same thing to her older brother and to her. It did not work. Her brother still knocked her over in their play.

Yet the same words came out of my mouth when I became a father and here was the same expression being used by my son to his son. I think we all hated it. It had been recognizably ineffective for four generations, running.

The phenomenon is real, but the explanation for it is missing. Why do we tend to do what has been done to us, to repeat the same relationships, to perpetuate our experiences? We do not know why or how, but I believe that the mechanism for it is in the brain. Certainly, we must remember what was done to us in order to repeat it. It is learned behavior.

People who have been tortured in childhood tend to become torturers of others. Not all actually perpetuate abuse but many must resist the temptation to do so. Psychiatrists used to call this identification with the aggressor—for example, a son's desire to emulate his powerful father. The abused become the abusers, picking on someone smaller and defenseless. This is the pecking order. It seems right to do what your parents did, even if you hated it. It feels good to be in control after years as the helpless victim.

It is amazing that most abused people can rise above the temptation to act as their parents did. They can make use of the wholesome influences that they have experienced and can integrate them into a pattern of action that stays within the law and even contributes importantly to the general good of society. In praising the resiliency of the human spirit, I am really praising the awesome plasticity of the human brain. But the brain must be healthy if the spirit is to overcome the handicap of an abusive environment. If the brain is damaged because of neurologic or mental disease, the temptations engendered by abuse may not be resistible.

CHAPTER 9

Anatomy of Evil

"Have you ever met a criminal who was just plain evil?" I have often been asked this question. It seems as if the medical-scientific approach to violence is exculpatory. If evil acts are a matter of brain function, is anyone really responsible?

The following analogy that I used before may help to explain my understanding of evil: when a composer conceives a symphony, the only way he or she can present it to the public is through an orchestra. Without these musicians, the audience can never know the composer's idea. If the performance is poor, the fault could lie with the composer's conception, or the orchestra, or both. If the orchestra has played on broken instruments, however, it is likely that instrument failure is a significant cause of the poor performance. Will is expressed by the brain. Violence can be the result of volition only, but if a brain is damaged, brain failure must be at least partly to blame.

It may be easier to understand medically how someone could lose control during an armed robbery, in war, in an argument, or even while disciplining a child than how someone could plan and carry out a cold-blooded murder. Brain damage could make it more difficult for a person to control his or her impulses, but what about serial murderers? They are mainly white men from middle-class homes. They do not carry the disabilities imposed by poverty, racism, and sexism, and they stalk and kill in a premeditated series of acts. Disease of the brain can impair a person's judgment and create abnormal, antisocial desires while decreasing the person's will to resist them. My examination of serial murderers has usually revealed evidence of mental illness, neurologic deficits, and abuse that is very much like what I encounter in "ordinary" murderers.

The social history of serial killers usually includes grotesque sexual abuse and/or severe physical childhood abuse. Many times the subject does not remember or denies the abuse, and other family members or acquaintances must provide the details or fragments of the story that hint at the extent of the abuse. However, family members may have strong motivations to hide the details of the killer's social history, especially when it is full of the grossest sexual abuse and/or severe physical abuse.

The triad of abuse, mental illness with paranoia, and neurologic deficit has been present in almost all the serial murderers I have examined. What I find puzzling is the obsessively repetitious features of the details of the crime in serial killing. Each killer has a particular modus operandi. This bespeaks some kind of perverse need that must be satisfied in a particular manner. The fact that most serial killers prove to be model prisoners may be because the prison environment prevents the subjects from repeating their crime in the manner of their choice. If the circumstances are not "right," there is insufficient motivation to kill. Serial killers are not brawlers with low thresholds to violence.

Serial killers usually manifest mood disorders such as depression or manic episodes with wide mood swings that last for weeks or months at a time. Murders are often committed during these episodes of depression or mania. In between, when the killers are neither depressed nor manic, they conduct themselves as mild-mannered and slightly ineffectual citizens.

Though brain damage is a frequent feature of serial killers, the damage seems less severe than in nonserial killings. Most "ordinary" killings are so stupid and vile and the perpetrators are so foolish and damaged that identification and arrest of the killer is fairly easy. But both types of killers may derive satisfaction from killing. The crime makes them into a "winner." It is probably more difficult for the police to identify and stop serial killers because they are more cunning and their motive for killing is less obvious, more idiosyncratic, and harder to identify. One big difference, therefore, between serial and nonserial killers is that the nonserial killer is caught before he can evolve into a serial murderer.

The perpetrator in the next case was a thirty-two-year-old white middle-class male whom I will call Whitney Post. He was quite typical of the serial murderers I have examined. He had been convicted of killing six prostitutes when I examined him. After killing, he amputated the feet of most of the women and the breasts of one. He killed all but one in a remote, wooded area. The last killing was in a busy part of a city, and this choice of location led to his apprehension.

I had been engaged by the defense to examine Whitney. No one had been able to derive from him what emotions he felt at the time of the murders. His description of what he had done had been emotionless, according to his attorney, and provided no clue as to his state of his mind at the time. Thus, the killings seemed to be "cold-blooded" to the prosecutor and the press, but the description of his behavior by near victims indicated that he was in the throes of an emotional upheaval at the time of the killings. There were some suggestions that Whitney was suffering from a period of depression, because all six killings had been carried out within a relatively short period of time, a few months. Four of the murders had actually been committed within a few weeks.

The prosecutor had done extensive preparation. His investigators had tracked down prostitutes in many states who had once worked in Whitney's city. The prosecutor was able to convince three prostitutes who had "dated" Whitney to testify against him and to tell the story of what he did and how he appeared to them, the survivors, during their encounters with the killer. The defense team that had engaged me provided me with a riveting transcript of their testimony.

Testimony of Miss Griffith

PROSECUTOR: "Miss Griffith, what is your age?"
MISS GRIFFITH: "Thirty-four."
PROSECUTOR: "What was your occupation?"

MISS GRIFFITH: "I was an active prostitute."

PROSECUTOR: "When did you begin that activity?"

MISS GRIFFITH: "Oh, a total of fifteen years ago, I imagine, fourteen, fifteen years ago."

PROSECUTOR: "Is the defendant, Whitney Post, someone you met in the course of your activity of prostitution?"

MISS GRIFFITH: "Yes, he was."

PROSECUTOR: "Are you a heroin addict?"

MISS GRIFFITH: "Not anymore."

PROSECUTOR: "Were you at the time?"

MISS GRIFFITH: "Yes, I was."

PROSECUTOR: "At the time you met the defendant, engaged in prostitution, can you describe or recall what your condition was in regard to your addition?"

MISS GRIFFITH: "I was withdrawing, I was sick."

PROSECUTOR: "Can you describe that for the jury, please, what that is like?"

MISS GRIFFITH: "Hot and cold sweats, stomach hurts, your legs hurt, sick to your stomach. It doesn't feel real good, like a real serious flu, I imagine."

PROSECUTOR: "Would you describe your first prostitution encounter with the defendant, Mr. Post?"

MISS GRIFFITH: "Okay. Like I said, my condition probably was not—I wasn't necessarily thinking as straight as I normally would. So, I went ahead and went with him. I asked him, you know, like, God, you know, what is your trip? He said he wanted to go into the woods, but I told him I didn't want to go as far out in the woods as he had asked."

PROSECUTOR: "Was there any price negotiated at that time?"

MISS GRIFFITH: "He told me he would give me $50."

PROSECUTOR: "Was that okay with you?"

MISS GRIFFITH: "Yeah."

PROSECUTOR: "Was there any discussion as to what his desire was with regard to you?"

MISS GRIFFITH: "He said he liked seeing a woman tied up. And I said well, I'm not going to let you tie me up. And so we agreed to that. I would simulate being tied up while he masturbated. And so I agreed to that. And we—he drove the truck to a retirement apartment building."

PROSECUTOR: "And is that where the date occurred?"

MISS GRIFFITH: "Uh-huh."

PROSECUTOR: "What happened at that time?"

MISS GRIFFITH: "Well, it went as he said. I laid face down on the seat, the passenger seat with no clothes on. And simulated being tied up while he masturbated."

PROSECUTOR: "Describe how you did that."

MISS GRIFFITH: "Well, I laid face down. I raised—I bent my knees and put my hands behind my back and around my feet.

PROSECUTOR: "What happened after that incident?"

MISS GRIFFITH: "He took me back to town and I got out of the truck."

PROSECUTOR: "Do you recall, Miss Griffith, still other instances in which you were with the defendant?"

MISS GRIFFITH: "Oh yeah."

PROSECUTOR: "In total, how many times to your recollection have you been with the defendant, Mr. Whitney Post, as a prostitute?"

MISS GRIFFITH: "At least fifteen, sixteen times."

PROSECUTOR: "Can you describe how frequently you would meet the defendant?"

MISS GRIFFITH: "Two, maybe three times a month. At least that I would see him driving around. I might have seen him like maybe twice a month or—and then I might not see him at all for maybe two months."

PROSECUTOR: "On your various encounters with the defendant, would there be any alcohol consumed?"

MISS GRIFFITH: "Yes, almost every time. He generally had the miniature airline bottles already in the truck. And we would stop and I would go in and get orange juice or whatever mix that—he always gave me my preference of what I wanted to drink, which was vodka. So, I would buy orange juice."

PROSECUTOR: "After your first encounter, you indicated you simulated being tied up; is that correct?"

MISS GRIFFITH: "Yes. That continued for a while. It wasn't like—I saw him the first time and then the second time I let him know that. That stayed pretty much the same routine for sometime."

PROSECUTOR: "Okay. Miss Griffith, what sort of areas would you go to on these dates?"

MISS GRIFFITH: "Oh, well, as time went on, first they were closer into town and then they gradually became, you know, longer, longer drives, further out into rural areas."

PROSECUTOR: "What went on? Did anything other than you simulating being bound and the defendant masturbating?"

MISS GRIFFITH: "No. You mean sexually?"

PROSECUTOR: "Sexually?"

MISS GRIFFITH: "No."

PROSECUTOR: "Did you ever have sexual intercourse?"

MISS GRIFFITH: "Never. He very, very seldom touched me. It was—he only touched my chest. He never really touched any place but my feet."

PROSECUTOR: "Can you describe his habit regarding the feet?"

MISS GRIFFITH: "Well, he requested that I arch my feet in a ballet position and hold it there."

PROSECUTOR: "How frequently did that occur?"

MISS GRIFFITH: "Every time that I had seen him."

PROSECUTOR: "Did he touch your feet with anything other than his hands as your dates progressed?"

MISS GRIFFITH: "Not until the last time I saw him. He requested that I go into the lady's room and wash my feet."

PROSECUTOR: "What happened after you washed your feet?"

MISS GRIFFITH: "Well, we went up in the hills and we had a regular, a usual date, in the process of doing that, he would like to bite my feet."

PROSECUTOR: "When, at what location did this biting of feet begin during the course of your dating him?"

MISS GRIFFITH: "Oh, it wasn't until, really until after I had let him start tying me up."

PROSECUTOR: "After you met the defendant the first time, was there a change from simulating being tied up to actually being tied up?"

MISS GRIFFITH: "Yes. It was about six, eight, nine months after I had first started going with him that I gradually started to consent to him tying me up with various things. First, he would tie me up loose enough to be able to get—to just release my own hands. And gradually that had changed too, to where I was totally submissive to him. I would be tied with nylons, bendable plastic coated wire, dog collars with belt buckles on them. Things I couldn't get out of."

PROSECUTOR: "In these instances, do you recall whether there was any pattern as to how the bonds would be removed when the date was over?"

MISS GRIFFITH: "Anything that was pliable to a knife was cut off of me. Like the nylons were cut off. There would be a dog collar around my wrists and another around my ankle with nylons tying the two together. And those nylons would be cut and the straps unbuckled."

PROSECUTOR: "Did you ever see knives in the car?"

MISS GRIFFITH: "He had a pocket knife in his pocket and several times had a box cutter or utility knife."

PROSECUTOR: "You indicated some biting of your feet during those instances; is that correct? Did that ever reach the point of pain?"

MISS GRIFFITH: "Yes. I would ask him not to bite so hard, and he eased up on me, didn't seem to be a problem."

PROSECUTOR: "Would he continue biting nonetheless?"

MISS GRIFFITH: "Oh yeah, after I felt that I had—I thought that I had got to know him, it wasn't a problem for me; I didn't have any problem with it at that time."

PROSECUTOR: "What times of day would these encounters happen?"

MISS GRIFFITH: "It could be any time of the day. I had seen him as early as ten, eleven o'clock in the morning. Afternoon, early evening, sometimes as late as two, three, four, five o'clock in the morning."

PROSECUTOR: "How is it that you were out at all these different times?"

MISS GRIFFITH: "I was a heroin addict; that's not unusual. If I was sick, I went to work. It didn't matter what time it was."

PROSECUTOR: "This pattern of tying you up and nibbling your feet, did that continue for awhile?"

MISS GRIFFITH: "Yes. The whole time I had seen him, basically, except he didn't start necessarily nibbling at my feet until I let him tie me up. But he always touched and fondled my feet. It was basically the same routine."

PROSECUTOR: "Prior to meeting the defendant, had you ever been to that rural area before?"

MISS GRIFFITH: "No."

PROSECUTOR: "Was this an unusually far trip in the course of prostitution to go?"

MISS GRIFFITH: "Very."

PROSECUTOR: "Why did you say you had no problem going with him to this location after a while?"

MISS GRIFFITH: "Well, in all fairness, what he was really asking me to do, considering what I was doing for a living, wasn't really that hard to have to deal with. There was no sex involved, so in the beginning, you know, it was fairly easy to do."

PROSECUTOR: "Did you have any reason to fear or distrust him at the time?"

MISS GRIFFITH: "No, not really. I mean, I didn't have any reason after I felt I had gotten to know him."

PROSECUTOR: "Were you paid on each occasion?"

MISS GRIFFITH: "Yes, every time. Most of the time he gave me $50. And if we stayed closer to town where I couldn't get all the way undressed or was undressed but it wasn't exactly the way he wanted it, he would only pay me $30. But toward the end, the last three times, two or three times, he had given me $40."

PROSECUTOR: "You indicated that there was a difference in these various encounters as to the state of undress of yourself. Would you describe that for the jury please?"

MISS GRIFFITH: "Well, from the first time that I got into the truck, the first thing that I did was take off my shoes. Shoes and socks, if I was wearing sweat socks. If I was wearing nylons, I'd leave them on."

PROSECUTOR: "Did the degree of undress that you engaged in change with regard to the location you were at?"

MISS GRIFFITH: "Well, I was always nude. Well, once in a great while I might, say, leave my top on, but generally I was nude."

PROSECUTOR: "Okay. The business with the feet, how frequently would that occur?"

MISS GRIFFITH: "Well, like I said, it was after I got to know him a little bit better. It was all the time except for the first two times I was with him. I mean, he would always touch my feet, but I mean the biting is what I'm talking about here."

PROSECUTOR: "Your feet were touched then on every occasion?"

MISS GRIFFITH: "Uh-huh."

PROSECUTOR: "On different occasions, were different devices used to tie the hands and feet?"

MISS GRIFFITH: "A couple of times he used a coat hanger, if that's what you mean. But I complained. It hurt the bones of my wrists and ankles. It was after that that he came up with the plastic coated wire."

PROSECUTOR: "On your last encounter with him, please describe what happened."

MISS GRIFFITH: "Well, we drove from where we were up into the hills. I laid down on the seat. And he tied my feet and hands. And tied them together. And then all of the sudden, for no reason at all, he started biting a little harder. It started off from the toes and biting down the balls of my feet, the arch of my feet, the heel and down my feet."

PROSECUTOR: "Were you bound at that time?"

MISS GRIFFITH: "I was hog-tied. Then when I asked him not to bite so hard, he didn't do it, he did the opposite. He kept biting harder. And instead of biting just my feet, he moved from my knees back down to my feet, biting my calves, both calves. It was intense, and the more I screamed, the harder he bit me."

PROSECUTOR: "You said you screamed, did you do anything else?"

MISS GRIFFITH: "I was screaming, squirming, you know, asking him why. I told him he could have his money back. I can remember getting hoarse. I can't remember if I couldn't hear myself scream anymore or if I was opening my mouth and nothing was coming out anymore."

PROSECUTOR: "Miss Griffith, other than biting your feet and calves, was there any biting elsewhere on your body?"

MISS GRIFFITH: "Yes, he bit me on the right side of my breast here."

PROSECUTOR: "On that occasion were any cutting instruments or weapons produced?"

MISS GRIFFITH: "Yes. He had something that was sharp enough to cut the bottom of my feet. It was hard for me to see it. It was at night. There was some light in the truck from the dashboard. . . . I couldn't see it clearly, but I thought it looked like the box cutter from where I was laying."

PROSECUTOR: "Describe what happened with the box cutter or whatever the instrument was."

MISS GRIFFITH: "He cut up the bottom of my feet. I don't know if he had it in his hand or on the floorboard of the truck. But he had bit my legs and feet for a period of time. And started cutting up the bottom of my feet. And told me that he was going to either cut up my breasts or my bottom and gave me the choice of which I preferred. I chose my butt."

PROSECUTOR: "Were you cut on your buttocks?"

MISS GRIFFITH: "No, he didn't do that."

PROSECUTOR: "You indicated you were screaming. Were you doing anything else at the time?"

MISS GRIFFITH: "I was trying to get away but it was—I was losing my energy doing that. I was tied up too tight. I remember I was wringing wet from trying to break loose. I remember totally losing my energy. And actually feeling myself separate. I remember being very quiet at one point where I couldn't say or do anything anymore. And that's when he stopped."

PROSECUTOR: "What happened then?"

MISS GRIFFITH: "He stopped and cut the nylon between the two devices binding my ankles and wrists, and turned me loose. He took the keys out of the ignition and stepped outside of the truck. I got dressed as fast as I could. I noticed my feet were bleeding, and I didn't have anything to stop the blood. I just put my shoes and socks on."

PROSECUTOR: "Miss Griffith, was there any conversation after the defendant stopped this biting on you?"

MISS GRIFFITH: "Well, I didn't say much of anything until I was sure we were going to leave. And then I asked him, 'Why did you do this after all this time? Why did you let me go?' And he said, 'I never killed anybody before in my life.' "

PROSECUTOR: "Where were you taken then?"

MISS GRIFFITH: "He took me all the way back to where I asked to be dropped. And then, as I was getting out of the truck, he kissed me on the cheek good-bye."

Testimony of Miss Ford

MISS FORD: "On my fifth encounter with him, he was playing with my feet, having me point them, and masturbating at the same time."

PROSECUTOR: "That's something you talked about early on too, right? Something about feet?"

MISS FORD: "He requested me to take my shoes off in the truck and I did. I put my bare feet on his side—the driver's side."

PROSECUTOR: "Do you remember what he said on your fifth date?"

MISS FORD: "Oh, about halfway out of town, he told me that he wanted to do things differently this time. And I asked him what he meant by different. And he said, 'I'm not going to hurt you or anything, I just wanted, instead of not touching you, I just want to be more physical.'"

PROSECUTOR: "What did you understand that to mean at that time?"

MISS FORD: "I assumed that he wanted sexual intercourse."

PROSECUTOR: "Up to this point, had there been any sexual intercourse?"

MISS FORD: "No."

PROSECUTOR: "What happened next?"

MISS FORD: "So we left the city and went to the rural area where we'd had our other dates. I started to get undressed. And he came up from behind me, which was normal, and started to put on the bondage devices. And soon as he cinched up the rubber bungee cords, they were too tight and they started cutting in. And before I could make any further statement that they were cutting off my circulation, they were too tight, he had strapped my ankles to my wrists."

PROSECUTOR: "Was this painful at the time?"

MISS FORD: "Yes. I told him no, please, that it was too tight and that the circulation was going out of my hands and my feet. When I told him that this circulation was going out and it was cutting into my wrists, he told me well, that's what he wanted to do, that was what he meant by when he said he wanted me more physical."

PROSECUTOR: "What happened next?"

MISS FORD: "He masturbated the same as he usually did. And when it was over, I thought he was going to untie me and he didn't. Then he left the truck and about fifteen minutes later or so, he came back to the truck and he jerked me up, took ahold of where my ankles and wrists were strapped together and threw me up onto the passenger—the seat across there and started biting on my breasts."

PROSECUTOR: "How did that feel?"

MISS FORD: "Excruciating."

PROSECUTOR: "What happened next?"

MISS FORD: "He was biting and tearing but not tearing in the literal sense—it would be if you clamped down on a fleshy part of your skin and tore, how it would break the top layer of skin. I told him to please stop, that is too rough, this is not right. I said all, a bunch of things like that. I sobbed and cried and said, 'Please stop' and I begged and he continued to do it. And the more that I begged and pleaded and sobbed for him to stop, the rougher that he got and the more the abuse continued."

PROSECUTOR: "What was his response to your pleading that this stop?"

MISS FORD: "He became more abusive. He was—the more I asked him to stop, the harder he bit me, the harder he tore at my breasts."

PROSECUTOR: "Did you at any time try to get out of that bondage?"

MISS FORD: "At that point, no."

PROSECUTOR: "What were you doing?"

MISS FORD: "I was just trying to cope with what was going on, trying to get him to quit by asking him."

PROSECUTOR: "What happened next?"

MISS FORD: "So, after this had gone on for about, I would say, fifteen to twenty minutes, I was to the point of almost being hysterical. And I was sobbing and crying so hard that apparently I was making too much noise. And he grabbed ahold of the back of my hair and jerked my head back. He put something up against my throat that I assume was a knife because it was cold and it was hard and I could feel an edge. And told me to be quiet or else I would really have something to cry about."

PROSECUTOR: "What did you do?"

MISS FORD: "I didn't say anything and I tried to stifle the sobs as much as I could, because I believed he was serious . . . that he would do some kind of bodily harm to me."

PROSECUTOR: "What happened next?"

MISS FORD: "He flipped me back onto the floorboard of the

truck. And then he got out of the truck and he left. I was looking out through the doors of the truck because both doors, my side was open and his side was open. And I was trying to get free. I was trying to work my arms out of the bondage devices but I couldn't because I couldn't feel anything in my hands or feet. And I started getting cramps in the back of my arms from having them in that position and there was no way I could do anything about it. He was gone for about an hour. And during that time, comprehending everything that was going on, I went into shock."

PROSECUTOR: "What do you recall happening with yourself? You say you went into shock."

MISS FORD: "A shut down of your mind and your emotions and everything else so that you could cope with what was going on and what had happened."

PROSECUTOR: "What happened next?"

MISS FORD: "He came back and he untied me and walked over to the edge of the lookout again while I got dressed. And I got dressed, he came back to the truck. We drove back down the road to the main road where the main road intersects. Just off to the side of that is a sandy beach, a small little beach area on the river. He pulled into there and walked down to the river and told me to come down there too. And I told him I couldn't."

PROSECUTOR: "Why couldn't you?"

MISS FORD: "I couldn't because of the way my legs and arms had been for so long and because of the emotional—experience that I had been through and everything. I just couldn't move. And then he came back to the truck and picked me up and he carried me down to the river. And—"

PROSECUTOR: "You mean he physically carried you?"

MISS FORD: "Yes. And he told me he was sorry, that he hadn't meant to get that carried away. He washed the blood off my wrists where the straps had finally cut into my wrists from being strapped that tight for so long and off my ankles and off my breasts. And then he picked me up and carried me back to the truck."

PROSECUTOR: "What happened next?"
MISS FORD: "He drove me back and dropped me off."

Testimony of Miss Elliott

PROSECUTOR: "What happened on your first date with the defendant?"
MISS ELLIOTT: "He asked me to remove my shoes. He wanted to see my feet. So I did. We agreed on the date and we left. He wanted to get out of the city and go to the country."
PROSECUTOR: "Did you get into the car?"
MISS ELLIOTT: "Yes. And right after I got in the truck, I took off my shoes 'cause he wanted to see the arches of my feet. He had me take—remove my pants and my shirt and turn around in the seat and put my feet up on the dash. And he rubbed my feet and masturbated."
PROSECUTOR: "What happened next?"
MISS ELLIOTT: "He got off. And I put my clothes back on and he brought me back into the city."
PROSECUTOR: "Did you have subsequent dates with the defendant?"
MISS ELLIOTT: "Only two."
PROSECUTOR: "Why? What happened on the last date?"
MISS ELLIOTT: "The third time he picked me up, I was dope sick. My bones were aching from heroin withdrawal. I got in the truck. I removed my shoes again; it was kind of a routine. I had already got it down pat after the first and second time. We turned off the road. It was easy for me to say yeah, you can tie me up now, because he didn't hurt me the first times. After he buckled me up, he got real aggressive."
PROSECUTOR: "Describe that buckling up that you just mentioned."
MISS ELLIOTT: "He buckled my hands with a strap, and my feet. Then, he had the buckle that came around that way. This time my face was in the seat. And he was hitting on my face as I turned it back and forth. I rolled my head."

PROSECUTOR: "Can you describe that? Was he hitting you with an open hand or fist?"

MISS ELLIOTT: "His fist. And he was hitting my face and lower back. I was crying. I was scared. The more fear I showed, the more aggressive he got."

PROSECUTOR: "Did he say anything?"

MISS ELLIOTT: "He was calling me a slut and said I was evil. There I was, screaming. He had gotten off and there was a truck light, a truck light up the road."

PROSECUTOR: "When you saw the truck lights, were you still buckled up?"

MISS ELLIOTT: "No. He unbuckled me. He got scared too. I got out of the truck naked and ran toward the lights. Then he caught up with me. I got in the truck and he took me to get my clothes."

PROSECUTOR: "Were you injured in any way?"

MISS ELLIOTT: "Other than the beating, my foot was stabbed."

PROSECUTOR: "Do you recall anything the defendant said at that time?"

MISS ELLIOTT: "He said he was going to take my feet."

PROSECUTOR: "When was the first time that you told someone about this stab or cut to your foot and the words you said, do you recall?"

MISS ELLIOTT: "To my foot? I believe it was when we were—I was with Detective Dolan."

PROSECUTOR: "Do you recall asking me a question?"

MISS ELLIOTT: "Yes."

PROSECUTOR: "What did you ask me?"

MISS ELLIOTT: "I asked if the women that were found dead, if they had their feet removed."

PROSECUTOR: "Did I answer you?"

MISS ELLIOTT: "You told me you couldn't answer me."

PROSECUTOR: "What did you say?"

MISS ELLIOTT: "I said, 'You already have.' "

Whitney Post came from a highly abnormal home in which severe and unusual punishments were delivered by his parents. The family moved two or three times each year, and every time the family moved to a place and Whitney's father found work, his mother would want to move again.

His father, Hubert, was a minister at a fundamentalist church. Bible reading and biblical exegesis provided by his father was applied to every area of the family's life. Respect for parents was a cardinal rule. As a result, Whitney did not like to speak to his mother, Rodina, because if he disagreed with her, it was considered "back talk." This would lead to the imprecation, "Go to your room and wait till your father comes home." Whitney described his room as a prison in which he would have to await punishment. His father, Hubert, would come in later—sometimes hours later. Meanwhile Whitney shivered in fear of the beating to come.

There was a ritual aspect to the punishment. When his father did arrive, he always positioned Whitney on his bed, on his stomach or kneeling, a rope or belt immobilizing his hands behind his back so that he would be unable to protect himself with his hands. Hubert pulled Whitney's pants down to his ankles, thereby also immobilizing his legs. Whitney was not allowed to squirm, remonstrate, or cry, as he would risk further lashes if he did. Whitney had to put his face in the pillows as he was forbidden to watch his father deliver the blows with a belt. He was struck ten to twenty times on his buttocks, back, thighs and the soles of his feet with the belt. These beatings lasted for several minutes and were delivered two to three times a week over a ten-year period when Whitney was five until he was fifteen.

In the course of investigating Whitney's background for the trial, I interviewed one of Whitney's younger adopted sisters, Michelle. I also spoke with his parents and his older biological sister, Susan. Michelle estimated that they were beaten every weekend from age three or four to about age twelve. She said, "Dad spanked us, and this always resulted in a frenzy. He was always in a frenzy. We were hit with the belt. Not a couple of blows, but many, rapidly delivered, for at least a minute. You would have to kneel down with your hands behind you. My brother's hands would be tied. Only the boys had to pull their pants down for Dad, but Mom used her hand or a fly swatter on the bare buttocks and the feet of both boys and girls. He did it to me too, but especially hard to Whitney. To hit somebody that many times you'd think he'd feel bad afterward. But Dad got into it—he enjoyed it." Sometimes Rodina came in to stop Hubert and would say, "I think that's enough." He would reply, "Don't get involved in my business." They would argue, but often the punishment ended at that point.

Michelle continued, "I used to pray he'd get into a car accident before returning home and get killed. The last time he did it I was twelve years old and the belt flew out of his hand and he picked it up by the leather part and used the buckle on me. It struck me on the lip and caused bleeding, and there's a permanent bump on my lip." She showed it to me. "After that, I said that if he ever did it again, I'd call the Welfare Department, and it stopped after that."

Whitney had said that his mother used a coat hanger when she beat him. She took a wire coat hanger, unwound it and made it long, and used it as a whip—a very vicious instrument. On other occasions, his father would discipline him and his siblings by pinching their arms to the point at which blood was brought to the surface and blood blisters and bruises developed. Another punishment that frightened him was when his father held him and shook him until his "arms became numb." While doing this, his father had "a crazy look in his eye." Other siblings agreed that Whitney was singled out for the most punishment.

The family culture was very disapproving of anything sexual. No function of sexual organs could be mentioned nor could any name of their anatomy be used. Yet there was considerable seductive behavior, often mingled with punishment. Massage was part of the family culture. No affection or hugging was done in the family, but massage provided an opportunity for some physical closeness, some pleasure, but also some pain that was often sexually charged. Hubert was the masseur for his children and others. When Whitney's oldest sister married, Hubert tried to initiate his new son-in-law into the family tradition. His son-in-law told the investigator that his first massage was his last. Hubert had a wild look in his eye. The massage was really a beating. The son-in-law described it as an extremely painful experience and something that was crazy. Some of Whitney's siblings said that they enjoyed the massages that Hubert provided them, yet they all agreed that the massages were often painful.

In my interview with an older sister, Susan, she said that her parents were very puritanical and very disapproving of sexuality and of the functions of those parts of the anatomy that are above the knees and below the waist. She said that in an attempt to prevent his sons from committing the "unpardonable sin" (masturbation), Hubert put hot sauce on their penises. She also acknowledged that her father's hands had wandered below her panty line and frightened her during several massages. Susan said she was frightened because she "didn't know what would happen." She refused to indicate that her father had ever been sexually abusive toward her. She also told me quite frankly at the beginning of her interview that she would not say anything that would cause pain to her parents. She was intent upon sparing them any further embarrassment.

She acknowledged that her mother characteristically walked around the house and the yard dressed in nothing but a slip, her nipples, buttocks and pubic hair visible through the thin garment. As children, both Susan and Whitney's other older sister, Sherry, were directed by their parents to walk around the house naked. They were both uncomfortable at the time with this, but they were not free to disobey.

All the children knew that Rodina refused to have sexual relations with Hubert because sex hurt her. They all knew that Hubert felt sexually unfulfilled. It seemed remarkable to me that such intimate details of the parents' relationship were the common knowledge of all the children in such a professedly puritanical home.

According to Michelle, after each beating, her father, Hubert, would leave the house and go outside alone. He would return a bit later. On one occasion when Hubert was home for a weekend he spanked several of the children viciously and left the house. Michelle secretly followed him out of the house. Hubert went to a shed in the back and she saw him masturbating. She was about seven years old at the time and went to her mother in amazement. "What is it if a man's pants are opened and his thing is out and it's in his hands and he's playing with it?" she asked her mother. Rodina replied, "Men are horrible. . . . All men are horrible and ugly and do disgusting things to women and you shouldn't let them do it to you or touch you."

I was anxious to find out about the origin of Whitney's interest in feet beyond his being beaten on his legs and feet by both his parents. He told me that it was his job to massage his mother's feet. I asked him to describe these sessions. He said that she would lie on her stomach in her bedroom, wearing only her slip, and ask him to rub her feet. He hated doing it and tried to do it for as short a time as possible, but she would request a more prolonged massage. She would moan and gasp softly as he rubbed her feet and he felt very uncomfortable massaging them, thinking it was more of her husband's role. He could not decline to participate as this would have been "back talk" and punishable by beating. Thus, he was helplessly drawn into a very sexually charged, incestuous situation. He found it very unpleasant to touch the skin of her feet.

Whitney told me, "My mother's feet were very soft, like a breast or my penis. There were no callouses. My hands and feet are like that too. Sometimes it is an embarrassment that my hands are so soft. My mother's skin is like mine. Her skin texture is different. Touching it had an effect on me. It was an unpleasant effect, not exactly like a woman's fear of touching a snake, but odd. 'Ooh.' " He shivered as he said this.

Whitney told me how he masturbated with prostitutes' feet. They were naked and would lie on their stomachs across his front seat or kneel on the front seat facing the back of his vehicle. He would straddle them with his back toward their head and would place his penis between their feet.

I told him that I thought there might have been a great deal of sexual abuse in his family that he might not have told me about because he did not remember it. I suggested that he subject himself to hypnosis, but he objected: "I fear hypnosis even more than death. If I do it and still get the death sentence, I'd have no identity any more. I would be branded, labeled, and I can't lay myself that naked to the world." He indicated that his family would be alienated. Dying without identity or loved ones to pay him tribute would "make me a nobody." He said that he, literally, would rather die than let others know the details of what was done to him by his parents.

But there is no doubt that there was sexual abuse in the home. Michelle, the younger adopted sister, told me that she witnessed their father sexually abuse another adopted sister, Teresa. Michelle said, "When I was about seven we lived in a shack. Behind the shack were woods on a hill. One of my brothers and I tried to tree a bobcat in the woods. We succeeded, and I went back to the shack to get something for the treed bobcat. I came down toward the house. My sister, Teresa, was there, bending over outside the playhouse and Dad was standing behind her. She was crying, and he had his pants down. I yelled. He stopped, pulled his clothes together and then spanked me. He said, 'No one will believe you.' He locked me in the playhouse. Then he came back and spanked me again."

When I told Hubert about Michelle's story, his response was disjointed. He said, "I don't believe it. Michelle is not truthful. I can't believe her. I can't believe hardly anything for sure. Some things she might not be intending to lie about. Perhaps I was urinating that time."

Michelle probably witnessed other sexual abuses by Hubert. She told me that when she was only three or four, she saw a girl who was about fourteen, with long black hair, tied to a tree in a field, facing the tree, naked. A man whom she thought might have been her father was standing behind the girl and beating her. Michelle was bothered by this image.

When I reviewed a videotape of a conversation the investigator had with Teresa, it was clear that she had been sexually molested on many occasions by their father. She described how he rubbed her back; how he rubbed between her legs; how he asked her to turn over on her back, while he massaged her. She could not remember everything that happened, but oftentimes she was sore the next day and couldn't remember how she became sore. Later in the interview, in tears, Teresa said that she had been sexually molested by her father many times with subsequent soreness of her breasts, genitalia, and inner thighs.

When I interviewed Hubert, he made it very clear to me that he would rather his son Whitney be executed than he be embarrassed by the "false accusations" made against him, the loving father. Hubert had very loose associations, a sign of a thought disorder, possibly schizophrenia or mania. When I told him that Whitney had told me that he would rather be executed than embarrass his father, Hubert wept and seemed relieved of a burden. When he regained his composure, his first statement was a complete non sequitur. "Music is a great way to turn off sexual thoughts. Music is a sexual release. My dead brother Lloyd said he didn't hear birds singing. Bob Bunnell is a tremendous bird whistler." He then expressed his great admiration for Eddie Duchin and Liberace and offered me two and a half hours of taped music if I would pay him four dollars.

When I asked him, "Are you saying that you did not do what Michelle and Teresa say that you did?" he answered, "If I had done something like that . . . I don't know what was in Whitney's head when he did that. I have never gone to bed without asking forgiveness for wrong doings during the day. Whitney was under the influence of the devil. He was not following the teachings of the church. He was off his rocker. I don't believe in hypnosis. No one hypnotized him or me. I never worked with the devil or operated a Ouija board. I'd have to be hypnotized by the devil. I don't think that way."

Rodina was also somewhat loose in her associations and began our conversation quite spontaneously by saying, "I was brought up differently from my husband. My parents were social, respectable people." She then went on to describe her father as abusive and violent. "My dad had a quick temper. He'd haul off and when I was a baby, I cried. He slapped me across my mouth and my tooth bled. He felt terrible, because he was so kindhearted. He was honest. A man worked for him on the roof and the man urinated up there. When he came down, dad socked him so hard that they had to call an ambulance to take him to a hospital. They thought he was dead. Once in a while the boys were spanked, and I would cry. My father was educated and had a good upbringing. My husband's family used English improperly and were uncouth. My husband was brought up like Topsy—poor manners. His mother was very stern, and my husband had a hard time in growing up." The family tree prepared by the investigator showed mental illness, neurologic illness (epilepsy and Tourette's syndrome), sexual perversion, and substance abuse to be prevalent on both sides of the family.

I tried to determine what Whitney's mental condition was at the time of the murders. There were strong indications of major depression. At the time of the murders he described himself as "run down." He would fall asleep in social situations, exhausted, yet he was not sleeping at night. He drank about twenty cups of coffee a day, which would wake him up, pick him up, and put him in a better mood. He was agitated but depressed. He procrastinated. He was short-tempered. He felt hopeless, helpless, and that his existence was worthless. "I was just spinning my wheels, not getting anywhere. My business was doing well, but I felt I had no future. I felt driven to date prostitutes. I wanted to stop. I felt guilty. I was in a frenzy. I was desperate. I felt totally lost." Whitney had lost thirty-three pounds during the period of the killings.

On the days he planned to see prostitutes, he couldn't do anything but think about prostitutes. He was, as he said, "compulsively addicted to dating prostitutes." Was this a form of Tourette's syndrome? Tourette's and obsessive compulsive disorder did run in Whitney's family.

His sex life with his wife of five years was sparse and not enjoyable. Typically, he would drink several ounces of vodka over a three-hour period on the nights that he was to date the prostitutes; he did not drink on other occasions. He estimated the frequency of his visits to prostitutes averaged three to four times a week. Once or twice a month, he would see prostitutes twice in a single day.

Before the murders began and during the period of the murders, he reported long periods of sleeplessness that lasted for weeks. He and his wife argued a lot. He was irritable. Normally, he would socialize with his sister Michelle, with whom he was close and who lived nearby. But during the period of the killings, Michelle felt uncomfortable with him—something about him seemed abnormal. She attributed this, at least in part, to rapid shifts from anger to contentment. Also, she said, he seemed to lack the capacity to reason, that he was incapable of adapting to changing plans, and his rigidity made it difficult for her to explain things to him. For example, if she had to postpone a plan to meet him for coffee because of her infant's needs, Whitney became unsettled, agitated, and angry. Patient explanation did not mollify him. This kind of rigidity can reflect obsessive thinking.

Whitney's earliest manifestation of depression may have been in grade school. There, Whitney's school performance had been substandard. He was hyperactive and did not pay attention properly. His behavior could have been labeled attention deficit hyperactivity disorder. The cause of his ADHD symptoms was not clear. At the time, teachers ascribed it to his difficulty in adjusting to different schools because his parents moved so frequently. Neurologic factors, obsessional thoughts, and depression also may have contributed to his poor academic performance.

Whitney had other symptoms of mental illness. He indicated that he had had a strong paranoid streak. He felt his teachers picked on him. He was always an outsider, a newcomer, lacking friends because of the family's frequent moves. Other children teased and criticized him for his ragged clothes and his unusual name. He was always "a lone wolf." He felt constantly blamed for things he did not do and was especially wounded by being belittled and humiliated by girls.

During our interview Whitney suddenly stopped and checked to see whether someone was listening at the door, thus indicating that his paranoia continued to the time of my visit. The glass door to the room in which we were located was closed and separated by an empty hallway from the main corridor. All was visible from where he and I were sitting. We were alone, and his concern was excessive.

He described an atypical mood disturbance with "natural highs" that lasted for about one hour, when he felt exhilarated and emotionally "good, powerful, and strong," as though he "could take on the world." He also described periods of dejection and depression that lasted for days and weeks, in which he felt discouraged and guilty, hated himself, and thought of harming himself. He cried often during these periods. Alcohol made him feel "loose, bold, brash," and he could say what he really wanted to and could express himself sexually. Under the influence of alcohol, he felt that his judgment was poor and his inhibitions were lacking. All the murders were committed when he had been drinking.

He also had a history of migrainous headaches, which lasted all day but were sometimes resolved if he could sleep. He had no major motor seizures, but he had rare déjà vu experiences, and his mother and sister told him he once sleepwalked into the kitchen, "sassed his mother," and then went back to sleep. The next day he had no recollection of the episode. It is difficult to know whether such lapses in conscious contact with his surroundings represented complex partial seizures; more likely, they represented dissociative spells.

On neurologic examination, I found that his grip strength was stronger on the right than the left; nonetheless he was left-handed, left-footed, and left-eyed. Usually, grip strength is slightly greater on the dominant side. His relative weakness of grip on the left suggested some deficit on the right side of his brain. Deep tendon reflexes were abnormally hyperactive on both sides of his body, and there were signs of frontal lobe dysfunction. I found primitive reflexes. Visual following was discontinuous and jerky, not smooth. He could not alternate strokes of his open hand, fist, and side of the hand (three-step Luria). He was only able to name seven words beginning with the letter *f* in one minute (normal is nine or more). He was able to read perfectly at a ninth-grade level but could not grasp the meaning of what he had read, indicating inattentiveness. All these pointed to frontal lobe deficits.

There was also some evidence of a thought disorder in that he interpreted the proverb "People who live in glass houses should not throw stones" bizarrely, yet concretely, saying it meant that "Their lives are fragile."

He had a number of scars, but the origin of most of them was unknown to him. He had a lengthwise scar on his right forearm, a one-inch scar in front of his right elbow, another scar above the left elbow, a deforming posttraumatic lump on his right inner leg, and scars on his right shoulder blade. Horizontal scars on his right and left flanks, his back, his buttocks, and behind his right knee and left knees were likely the result of beatings.

My finding of frontal dysfunction was confirmed by psychological testing. On the Category Test of the Halstead-Reitan Battery, he made sixty-four errors (normal is up to fifty). On the Wisconsin Card Sort Test he was only able to switch sets three times, whereas normal adults can do this four to six times. While the EEG provided no support for a diagnosis of epilepsy, a SPECT scan demonstrated abnormality in the frontal lobes that, along with the psychological tests, supported my clinical observation that there was frontal dysfunction.

My evaluation of his history indicated that spells of depression made irresistible the expression of impulses that had been conditioned by a decade of abuse as a child. In the months just prior to the period in which the murders were committed, Whitney had been a great success in his repair shop. His business was going well but he began to feel that he could not keep up with it. Dismal, self-deprecatory feelings focused on his business and his marriage and rattled around in his head, obsessively. His sense of victimization increased along with his depression. His perverse urges too became stronger, and he would go to the city to seek prostitutes, sometimes more than once a day. He wanted them not only for sex, but for domination. His ability to control prostitutes relieved his own sense of victimization, but his business began to falter and his marriage started to dissolve.

He was intensely ambivalent about murdering the prostitutes. He controlled his urge to kill in many of his outings and would debate with himself, sometimes for hours, while the prostitutes lay bound and naked in his vehicle. Their terror fed his desire to torture. Ultimately, he killed six women. Those who could dissociate, "separate," "become comatose," and cease demonstrating their terror lived, as he was able to marshal his controls. He even kissed Miss Griffith goodbye and apologized. He carried Miss Ford to wash away the blood stains his rough treatment had caused. But I imagine that those prostitutes who could not dissociate, who continued to squirm, to plead, and to cry out in uncontrolled terror only gripped his darkest passions and ignited a frenzy. Their testimony of terror could only be provided through the efforts of the medical examiner.

On the night of the final murder, he was intoxicated, or at least heavily under the influence of alcohol, as he had been at the time of the other murders. He felt despondent. He encountered and engaged a prostitute, then murdered her in full view of bystanders. Unlike the previous murders he committed in a remote area, this last killing seemed to demonstrate his decreasing capacity to control himself.

The anatomy of evil in the case of Whitney Post was similar to that of many other nonserial killings. Depression and neurologic deficits in the frontal lobes, accentuated by alcohol, forced the expression of the blackest urges that had been engendered by his perverted parents.

The prevailing motivation of the dozen or so serial killers I have examined is to control and dominate others. Victims are humiliated, beaten, tortured, dismembered, and occasionally consumed to exalt the sense of power of the perpetrator. In killing, the criminal shakes off his sense of victimization. Because periods of depression maximize the sense of victimization, they are dangerous. Mania releases the subject from the constraints that his normally functioning brain would provide. Thus, affective disorder with depression and bipolar disorder are common triggers for serial murder. Between episodes of affective disorder, most serial murderers are relatively benign citizens.

Most people who have undergone sexually damaging childhoods do not become murderers, let alone serial killers, and most serial murderers do not kill every day. Most of their killings occur in spates, usually within a few weeks of one another. This grouping of murders suggests periods of depression or mania. Neurologic impairment also damages the barriers that restrain violent actions. Intoxication intensifies and creates neurologic impairment. Alcohol is the most commonly used intoxicant. Cocaine and methamphetamine are the most dangerous as they regularly produce a manic state with paranoia, and withdrawal from these drugs often induces depression. Intoxication and neurologic impairment really produce the same effect on behavior; both disinhibit.

The methodical, compulsive aspect of serial killers and the sexual content of their motivation and actions set them apart from most "ordinary" killers. Each serial killer has a different method of operation that incorporates into the crime details that have idiosyncratic, sexual meaning. For example, Whitney Post fondled and amputated feet. This seemed related to his mother's demand that he massage her feet. She coerced him to perform this act in a sexually charged situation. William Bonin raped male hitchhikers, hog-tied them, and squeezed their genitals as he garroted them. Joel Rivkin dismembered the bodies of dead prostitutes in his parent's home with an X-Acto set. Jeffery Dahmer saved and ate portions of his victim's bodies. Douglas Clark kept the fully made-up head of one of his victims in his refrigerator. It is difficult to ferret out the relationship of life experience to each of these details, but the horror of their acts probably is an echo of the ghastliness of the murderers' own childhood experience.

This compulsive-obsessive quality of serial killers may prevent them from being violent when they cannot "do it the right way"—for example, when they are incarcerated. The need for the right time, place, and victim both stamps these acts as bizarre and protects fellow inmates from the serial killer's vulnerabilities to violent action. In this way they differ from "ordinary" killers, but they are not normal people who suddenly become serial killers because they are evil. Like "ordinary" killers, their killings are largely the result of abuse, mental illness, and neurologic damage.

Another symptom that "ordinary" and serial killers often share is the ability to dissociate. Occasionally this develops into a full-blown dissociative identity disorder.

CHAPTER 10

Did You Ever Hear a Baby Cry?

In the first three years of life, the brain undergoes tremendous enlargement. At birth, the brain weighs 400 grams. Its growth does not end until the beginning of the third decade, when it weighs 1,100 grams. Most of the growth occurs in the first three years of life. A major factor in this enormous growth is the development of connections (synapses) between nerve cells. During this period, the brain's hardwiring is established. The psychosocial environment is largely responsible for the maturation of the brain during this period. The impact of the environment through the senses—vision, hearing, touch, and so on—is dramatic and specific, not merely influencing the general direction of development but actually affecting how the intricate circuitry of the brain is wired. Synapses that are used during development are the ones that last; unused synapses disappear or become ineffective.[1]

If a child does not experience positive stimuli such as love and affection and is instead neglected and even tortured, parts of the brain may never develop properly. Permanent changes in behavior and thought may result. The behavioral effects of the lack of maternal nurturing in infants was first reported by Rene Spitz, Sally Provence, and others. The Harlows studied monkeys with regard to the lack of maternal nurturing and reported permanent changes from deprivation in the first months of life.[2] These considerations led to the nationwide replacement of orphanages with the current foster care system. But Eastern Europe, especially Romania, has a severe social problem posed by thousands of children living in the streets who had been socially and emotionally devastated by institutionalization early in life.[3] The changes in the brains of abused children that have recently been described are likely not just a minor anatomical difference. They may be the physical manifestations of a psychosocial catastrophe that has produced permanent neurologic deformities.

What if the tears of an infant are totally ignored, if his signals for help are consistently misread or disregarded, if his mother punishes his persistent crying by striking him, slapping him with a belt, or burning him with a cigarette. If his cries of pain and rage are ignored or precipitate more punishment, the child will learn not to cry. He will lie inert, withdrawn, and appear to be developmentally slow. Yet he may be hypervigilant and alert, looking for environmental clues that signal danger, trying to prepare himself for futher punishment.

Abused toddlers often have difficulty relating to other children and adults. They may be socially isolated, suspicious of the intentions of strangers, hyperactive and alert, destructive of property, and often act aggressively. The development of motor skills may not be delayed, but abused toddlers are socially immature and slow to learn. Older abused children enjoy causing pain to animals and are unsympathetic to peers who are crying for help. Fighting, persistent bed-wetting, destruction of property, and setting fires are common symptoms of abused children, as is attention deficit disorder with oppositional defiant disorder and conduct disorder. Although there is a strong correlation of these symptoms with abuse, not everyone who is abused has these symptoms, nor have all children with these symptoms been abused. Brain damage and psychiatric illness can produce similar symptoms. Children with all three factors—abuse, brain damage, and mental illness—are especially vulnerable to criminality and violence in later life. Mentally ill and neurologically damaged children behave badly and become magnets for parents who beat their children for misbehavior. Abuse can cause physical damage to the brain. By affecting behavior, brain damage can also elicit abuse, a vicious cycle.

Because abuse is such a potentially important factor in the

development of the brain, it is essential for medical, social, and education personnel who are investigating abnormalities of behavior and cognition to determine whether an individual has been abused. Unfortunately, this determination cannot be made easily. Abused children and adults deny that they have been abused for at least two reasons. First, they simply do not remember the abuse, having blotted out those experiences from their conscious memory. Second, they prefer to think of their parents' behavior as normal, not "abusive." When I simply asked violent delinquents and murderers if they had been abused, two-thirds said "No." If I had not investigated futher, I would have thought that abuse was not especially common among violent delinquents or among murderers on death row. Abused children who are in the custody of abusing adults also realistically fear reprisals for "telling."

Victims of serious psychic trauma often do not remember it. This forgetfulness is the same as that seen in people with post-traumatic stress disorder who have difficulty recalling specific autobiographical memories. Because large segments of time can be forgotten, even years, this memory loss can have a major impact on behavior.

Social development depends on a child's ability to learn how to solve life's problems effectively. What children learn from their mistakes, failures, and successes guides their future actions. Our ability to solve problems effectively depends on the ability to remember how we solved similar problems in the past. We must know what succeeded and what failed. During development, the neural networks underlying the various possible strategies for solving problems become part of the intrinsic mechanism of the brain.

When children are put in situations of physical danger, from which there is no escape, when they are tortured and threatened by their parents, or by the only people who are in a position to protect them, they may involuntarily forget what happened to them. The erasure of substantial periods of childhood from conscious memory has a good and a bad side. Though it may allow a child to emotionally survive the abuse, it causes several long-term problems. Not only does the child miss lessons concerning social responsiveness, but the process of erasure itself becomes a coping mechanism. The automatic brain reflex for many abused children is not to solve problems but to dissociate. In other words, the learned pattern of disengagement does not simply mean that adults cannot remember portions of their childhoods; it often means they have not developed critical life skills such as how to manage difficulties with interpersonal relationships.

How do you deal with somebody with whom you are having an argument? Constructive options include explaining your point of view and apologizing. The choices for dealing with disagreements are learned, and the most important teachers are parents. If abused children can only deal with parental conflict by disengaging themselves either physically or psychically, then they do not learn how to solve critical social difficulties constructively. Instead, they may become inappropriately aggressive in response to a disagreement.

When feeling threatened by an outsider, an abused person may dissociate. He "disappears." He may actually develop the fantasy that another person experienced the punishments—someone else who will strike back in vengeance at the outsider who threatens him. Obviously, there is only one person—the abused person with a brain that has been importantly damaged by experience. In such cases, there is often a "nice personality," or host. The alternate personality regards the host as a wimp. The alternate is a violent, nasty avenger who takes risks, gets into serious trouble, and attacks other people whom the host does not like. The host personality frequently does not recall the actions of the avenging alternate personality. Some people have evolved many different personalities. In this extreme form, the tendency to disengage from problems rather than to really solve them results in dissociative identity disorder (multiple personality disorder). People who use this mechanism become erratic and variable in their morals and their reactions in different situations, depending on the personality that takes over. Some of the different personalities may even represent the opposite gender.

Uncovering this tangled complex of behavior, motivations, identities, and forgotten or half-forgotten experiences is not easy. The physician cannot depend solely on what the patient remembers and must carefully pose questions that are not threatening or belittling in order to discuss what the person does remember. The suggestion that one's parents were torturers is often perceived as very threatening.

The concept that a personality disorder may cause violent behavior is not often extended to dissociative identity disorder (DID)—the modern term for multiple personality disorder. Prosecutors do not like DID because its diagnosis tends to be exculpatory and/or mitigating. The life experiences that cause DID are so bizarre and horrifying that most prosecutors do not wish to acknowledge them, lest the court sympathize with the defendant. Many doctors, including many psychologists, do not believe that DID exists, even though it is a diagnosis sanctioned by *Diagnostic and Statistical Manual of Mental Disorders,* Fourth Edition (DSM-IV), the diagnostic "bible" of American psychiatry. Doctors who do not accept DID ascribe multiple personalities to lies, suggestion, and imagination. Defense attorneys are reluctant to introduce the diagnosis because it seems so much like literary fiction that they are afraid jurors and judges may not take it seriously.

I first met Leo Murphy when he was on death row in Arizona, convicted of felony murder. He and another man had been in the process of burglarizing an empty house when the owner returned and found him. The owner was armed, pulled a gun, and shot at Leo, missing him. Leo had not been armed when he entered the house, but another loaded pistol belonging to the owner of the house was lying on a shelf. Leo picked it up and used it to shoot and kill the owner.

Leo's coconspirator turned Leo in to the police to obtain a milder sentence. On a tip from the coconspirator, the police searched Leo's belongings and found stolen items from the home of the murdered man. The police obtained a detailed and signed confession from Leo, but he later claimed that he did not commit the murder. Leo said he had no recollection of having shot the man, having entered his house, or having been involved in the robbery. He also denied admitting his guilt to the police and signing a confession.

Leo was a friendly, articulate Caucasian man in his late forties. He spoke in a forthright manner and came across as a likable person. He described many peculiar blank episodes in his life lasting from minutes to hours and sometimes for days. Many of his memory lapses involved burglaries. He freely admitted to being a "cat burglar" who operated during the day in affluent neighborhoods. He said quite frankly that he chose expensive homes in good neighborhoods because these would be insured and the owners would not suffer any major loss. He took jewelry and occasionally cameras and a pistol if he could find it. These he would sell to a "fence."

His standard procedure was to knock on the door of a home; if someone responded, he would not burglarize the house. He never entered a house armed or one that he knew to be inhabited at the time. He said that he had performed up to five burglaries a day and his all-time personal record was performing four in one hour. Leo freely admitted these activities to me.

He described a sense of exhilaration at being in the homes of people, a sense of power, of freedom, and of release. On one occasion he accidentally entered the home of someone who was sleeping. He crawled very quietly around the sleeping individual and stole things from the bedroom. He described the empowerment he felt then as one of the greatest experiences of his life, but he did no harm to the inhabitant of the house.

Leo was suspected of having performed one violent crime prior to this: the rape and murder of an elderly woman. The evidence against him was so weak that he had not been charged, but both the police and his own lawyer suspected that he had been involved in that crime. He denied it and denied any knowledge of the dead woman.

Sometimes Leo would find himself standing in a house with a pile of things waiting to be taken out that had been gathered, clearly by himself, but he did not remember breaking into the house or collecting those materials. On some occasions following burglaries, among the stolen possessions in his van he found women's underclothing, panties and bras, for which he had no explanation. Often the stolen underpants were soiled. He did not know how they came into his possession and did not remember ever taking them or even wanting to take them. As a youngster, the records showed, he had been caught stealing things for which he had no need, including women's soiled undergarments, and about which he could offer no explanation.

Leo told me that he often smelled the odor of soiled undergarments, but which no one else was able to smell. These episodes of olfactory hallucinations lasted for minutes to hours and had occurred intermittently over the course of his entire life. He also experienced frequent déjà vu and had out-of-body experiences, which were often associated with migrainous headaches that lasted for hours and sometimes days. These began during adolescence. He had never had a witnessed epileptic seizure.

Spontaneously, he said that he believed in "astral projection"—that he could leave his body and go anywhere. One of the places he would go would be to a quiet, wooded area to get away from things. He would go to this place especially when he was feeling bad.

Dorothy Lewis, a psychiatrist with extensive experience with DID, and I both examined Leo. We have often worked on the same cases, but this was one of the rare occasions when we sat together with a subject. I had started the interview, but when Leo mentioned "astral projection," Dr. Lewis and I agreed that she should take over. "Astral projection" was a term we had encountered several times before, invoked by patients to describe dissociative states.

Leo said that he was an avid reader of novels and that he would put himself in the book: "I'm there when it is happening." The same thing occurred when he watched movies or television. He felt as though he were actually in the story. He tended to identify with the powerful figure, usually the hero, sometimes the heroine, and sometimes the villain in the piece. The latter was especially true of horror movies, which he especially enjoyed.

Leo said he very rarely used drugs, but on the one or two occasions on which he snorted cocaine it gave him feelings of energy, alertness, a "good, high feeling," and made him feel "invulnerable." He claimed to have had such experiences without using drugs as well. These experiences resembled manic episodes and were associated with his criminal activity, such as burglarizing houses: "When I go into a house, I have a high, adrenaline feeling. I hear everything. If there is a fly in the room, I know it. I become one with everything inside. I feel as if I am floating on air. It's a good feeling. There is nothing in the world I can't do. It's a rush."

"Natural highs," defined as unrealistically euphoric feelings that occur spontaneously, have much in common with cocaine and amphetamine effects. During periods of mania, patients with bipolar affective disorder describe racing thoughts and feelings of empowerment, capability, invulnerability, energy, and release. Manic patients may gamble, be promiscuous, spend money unwisely, be unable to sleep, demonstrate rages, endanger themselves and others, and abuse drugs and alcohol. Episodes of mania can last for weeks, but brain-damaged individuals with mania often experience briefer episodes of great intensity.

Asked what he did with the money from his burglaries, Leo responded, "I'm like a kid. I buy things I don't need. I've used the money for bills but also take the kids to an amusement park, dinner with my wife in a restaurant, and I bought a BB gun. I call my (biological) mom and speak for half an hour. I have so much energy my mind never stops, it races. I can't keep a thought very long before I go on to something else. I am distractible, and I can't sit very long." He had insatiable sexual urges, masturbating up to four times a day.

Leo also had periods of depression when he felt sad, hopeless, helpless, worthless, and was unable to sleep for more than fifteen minutes at a time and for not more than four hours in a twenty-four-hour period. He wouldn't bathe for days at a time and once went for a whole month without bathing. He wouldn't change his clothes, had no appetite for food, lost his libido, and masturbated only about once every two weeks and cried daily. As he told us about his depressions he wept. There had been three recorded suicide attempts when he overdosed on medication.

His early school history was consistent with the diagnosis of attention deficit hyperactivity disorder with conduct disorder and learning disorders. He was especially poor in arithmetic and spelling and had never completed the tenth grade. His early academic and behavior problems had attracted the attention of school authorities and social agencies. Their reams of reports filled with data about his family life, upbringing, and symptoms were the source of most of what we learned about Leo.

His biological mother was an alcoholic who drank heavily during her pregnancy with him. Although his head was of normal size, I recall that the thin vermillion border of his upper lip and his shallow philtrum (the vertical indentation between the upper lip and the nose) were signs of fetal alcohol effect. Abnormalities in his performance at school and on psychological tests might very well have been related to such prenatal factors.

On physical examination there was unequivocal evidence of neurologic abnormality that indicated that both sides of his brain, especially the frontal lobes, were dysfunctional, more so in the right hemisphere of the brain. Both EEG and PET scan were consistent with bifrontal and right temporal damage. The MRI scan of the brain showed focal atrophy of the right temporal lobe.

There were many scars over his body, including multiple horizontal scars on his back from whippings with a wire. In addition, he repeatedly pulled out his own toenails and several of his toenail beds were healing. He could not say why he had pulled out his toenails and did not specifically remember ever having done so. According to his records, he had claimed that voices told him to remove his toenails.

Leo's uncle had told an investigator that "Leo has a dozen distinct personalities." Everyone who knew him well was aware of these different personalities and that he did not always recall conversations or events. These observations of others about him were first made when he was in grade school. The blank episodes, the out-of-body experiences, and the "astral projection" described by Leo definitely pointed to dissociative episodes and possibly to a multiple personality disorder.

Such symptoms are most often encountered in people with a rather dramatic history of physical and sexual abuse, which proved to be true in this case, too. Leo's family and social history detailing physical and sexual abuse had been obtained by an investigator long before we saw Leo. His biological mother was said to be a prostitute who used Leo as her lookout during his first decade of life. As a child, he was not highly regarded by his biological mother, though they had reconciled in his adulthood. She physically and verbally abused him. For instance, he said his birth mother had told him he was a "piece of shit" and his birth had, for her, been like having a bowel movement. When Leo's parents divorced, his father, a brutal truck driver who was usually away from home, obtained custody of his son. His second wife, Leo's stepmother, can most charitably be described as a perverted sadist.

The stepmother hung Leo by his thumbs; pulled his thumbs out of their sockets as part of punishment; burned his hands on the stove; forced him to smoke an entire pack of cigarettes, burning his lips and fingers, at age ten; repeatedly beat him with a wire brush and a belt, prompting neighbors to call the police; pushed him down a flight of stairs to the basement with his hands tied behind his back; chained him to a pole in the basement for days dressed only in his underwear without allowing him to use the bathroom; and locked him in a small closet for days. After such excessive periods of confinement she would rub this own fecal matter in his face.

He also suffered obscene sexual abuse at her hands. When he was ten his stepmother told him she wanted him to look like and be a girl. She took a string and tied it to his penis, pulled it tightly in back of him so that it could not be seen from the front, and tied the string to his waist and told him he looked better that way. She gave him enemas using Vaseline and inserted tampons—at first just clean tampons and then later her own soiled, bloody tampons—into his anus. She made him insert tampons into her vagina and forced him to fondle her and perform cunnilingus until she climaxed. The only experience of tenderness to which he had access was in such a setting. He described her as being gentle and nice to him while he was sexually stimulating her.

She would also stimulate him manually and orally but would not allow him to reach an orgasm once he attained sexual maturity. When he was in his early teens, if he did reach an orgasm during this sex play, she would take the semen and smear it on his face and would angrily punish him. In order to prevent him from reaching a climax she would squeeze his testicles, painfully. The factual basis for most of these assertions was supported by members of Leo's family, former neighbors, a psychiatrist, and a psychologist. The psychologist had spoken to Leo's stepmother, who admitted her profound and extreme physical abuse of Leo. The family reluctantly admitted that they knew of some of these things but were so accustomed to perversion, incest, physical beatings, and alcohol and drug abuse in their own lives and homes and were so overwhelmed by mental illness and depravity that they could not or would not intervene effectively.

When Dr. Lewis asked Leo what he thought of when he was being punished by his stepmother, Shirley, he said, "When I was being punished I could go into the woods by a creek. In my head. I felt safe there. When I'm depressed or having a bad day or having an argument, I can go there now." This was what he called "astral projection," the ability to leave his physical environment and enter an imaginary world.

I knew that Dr. Lewis had for many years managed a clinic for DID children at Bellevue Hospital in New York, but I had never seen her interview a patient with DID. I was astounded by some of the questions she had learned to ask.

Dr. Lewis asked, "Did you ever feel as if bad things were happening to someone else?"

He responded, "When Shirley beat us my brother ran away but I watched her face. I saw the hatred in her eyes. I saw the blows on myself. I saw myself cringe."

"Did you ever not feel it?"

There was a long pause. "No," he said as he began to cry, silently. Tears streamed down his face. "I saw what she did. It was evil, how she used to beat me, laugh at me. It got to the point where I hated myself and I couldn't do anything about it." He turned his face and looked at the floor. He spoke almost under his breath, "I hated her so much."

Dr. Lewis asked, "Do you have a friend you can talk to?"

"I do have a friend I can talk to. It sounds like a woman's voice."

"When did it first come?"

"It first came when Shirley started beating me. She (the voice) told me everything would be okay later. She was caring. I was nine and a half or ten when she first came. There were other voices, but I never paid attention to any of the other voices but hers. I have heard a lady's voice in prison that says everything will be okay and that you are trying to help me."

Up to this point in the interview, I felt that we were on familiar ground, but Dr. Lewis's next question stunned me. Dr. Lewis asked, "Did you ever hear a baby cry?"

"I heard a baby cry when Shirley beat me. There was a baby crying or cooing in the background." I wondered who the baby was. Later Dr. Lewis told me that many patients with DID hear a baby crying when they are being tortured; probably the "baby" is themselves.

Dr. Lewis asked if the lady's voice in his head had a name, and Leo indicated that the name was "Terry."

Dr. Lewis continued, "Who taught you about sex?"

Leo answered, "Shirley taught me about sex. She put me in a closet, and I peed in my pants. She beat me until I peed or crapped in my pants. She'd call me a baby and put a dish towel on me like a diaper. I'd cry out to go to the bathroom or to get something to eat and if I crapped she'd smear it on me. She dressed me in girls' clothes. She said I'd make a better wife than a husband. She'd call me 'honey.' Later on I had bleeding from my ass when she pushed tampons up my butt. She'd have me insert it in her and then she'd take it out of her and put it in me. The first time she did it without Vaseline. She'd put her bloody tampons in me, sometimes two or three at once. She gave me enemas with animal blood because she was a butcher and she'd get it from the slaughterhouse. She did that so I'd be like a girl having a period. She'd give me hot, soapy water enemas and then make me take a shower with her. Afterwards she'd take me into the bedroom and make me . . ."

At this point he stopped talking, sat silently for a minute or two with tears streaming down his cheeks. "After she made me 'clean' she took me into the bedroom and made me perform. She said I had to be cleansed and she would take a string and tie my dick with it around my waist. She dressed me in panties and made me go down on her and play with her. She used to tell me to call her 'Terry.' " I found it remarkable that Leo's figment of imagination, Terry, had the same name as the abusive Terry, who may have been an alternate personality of this cruel stepmother, Shirley. Two people with DID?

Leo continued, "Shirley wanted me to call her 'Terry' when I did the things she wanted. Terry was nice to me, but other times she wanted to be called 'Mother.' She smacked me when I called her Shirley. She never let me come when I was older. After I made her come she'd send me to the toilet and told me to jerk off in her dirty underpants."

Dr. Lewis asked, "Have people ever told you that you were like two different people?"

He said, "People have told me that at times I am more sensitive than at other times, that I care about them and have a kind heart. Other times they say I am as cold as ice, a different person."

"What names have people called you?" Dr. Lewis asked.

"Well some people call me Tony. I told them I don't know who they are. That's not my name. I walk away." It was bizarre. I had the feeling that we had entered a kind of malignant wonderland. The people who called Leo "Tony" were also figments of his imagination.

"What other names have people called you?"

"David."

"What is Tony like?" Dr. Lewis was questioning Leo, the host presumably, about his alternate personalities.

"He is a loving person."

"Is there an angry one, an angry voice?" This was a vital question that Dr. Lewis posed. The personality that takes the abuse and the pain to spare the host is often an angry, vindictive personality who commits violent crimes of which the host is unaware. This angry and aggressive personality is brave and fearless and often contemptuously dismisses the host as a "wimp."

He answered, "She won't let me tell." We were at an impasse. A presumably controlling personality had just told Leo not to speak about "the angry one."

Dr. Lewis asked, "Leo, go to your special place where you feel safe and relax there. While you're there, I want to talk to the person in charge. I want to talk to the strongest. Who is the brave one? What did you do for him?" Dr. Lewis had not hypnotized Leo but had entered his imaginary world almost as an adult plays house or school with a child.

Leo answered in a soft, higher-pitched feminine voice, "I took him away."

Dr. Lewis shared a glance with me as to say, "Now we're getting somewhere." Dr. Lewis asked, "How old was he when you took him away?"

"He was a boy about ten or eleven. I've protected him since then."

"What did she do to him?"

"She was mistreating him, beating him, treating him bad, treating him like shit. Locking him in closets."

"What do you know that he doesn't?"

"She tied him to a pole and beat him with a wire brush handle. She's hurting him (he began to grimace). She's smacking his hands with a ruler. She ties him up and throws him downstairs. His hands are behind his back. He is wearing girls' panties. He is dressed like a little girl and she calls him 'asshole.' She threw him down the stairs and cut his lip. His mouth is bleeding. He's scared."

"Does he go to a hospital?"

"No, not to a hospital. She leaves him down there."

"What do you do?"

"I tell him it will pass. I tell him he'll be okay. I help him to his bedroom." The voice added with emphasis, "She should die."

"Did you ever try to kill her?"

"No, I was afraid for Leo, that he'd be hurt."

"You sound very gentle and you have been very helpful. Please write your name." Dr. Lewis extended a legal pad and a pen. Leo wrote "Terry."

"Terry, how good to meet you. Why didn't you tell us? We like Leo and are trying to help him just as you are. When did you first meet him?"

"Shirley started beating him when I met him. I live with Leo and go with Leo. I always try to be with him."

"Do you know Tony and David?"

"Yes."

"What is David like?"

"He's shy, a loner. Tony is outgoing. He likes people and he is trusting."

"Does he like us?"

"He hasn't made up his mind."

"Who is the boss?"

Leo looked indecisive. He paused and then said in the same high-pitched, measured, feminine, voice, "I am."

Dr. Lewis, relying on her vast experience with DID immediately picked up on the uncertainty Leo had just demonstrated. "No, you hesitated. Who gives you orders?"

At this, Leo threw an extremely nasty, threatening look at Dr. Lewis.

"I can't tell you."

"Why can't you tell me?"

"They say 'no.' She won't let me."

"Is a woman the boss?"

"Yes."

"Does she boss you?"

"Yeah, sometimes."

"Who is the one who gave me a cold look that scared me?"

"Lisa."

"Is she the boss?"

He nodded. "We all listen to her."

"How old is Lisa?"

"Lisa is older. She's in her fifties." All the childhood sex abuse and cross-dressing had produced female identities for Leo. Personalities of both sexes are common in patients with DID.

"Leo told us he doesn't remember many things. He doesn't remember going into some of the houses, but wakes up in a house and finds himself there with some stolen goods in his possession."

"Terry" continued, "David and Tony do this. They're trying to get him in trouble."

"Why?"

"They're nice, but they're screwing with his brain. I try to keep him out of trouble but sometimes he won't listen. I have to fight with the guys. I told them to leave him alone. They want to keep hurting him by keeping him in trouble. They don't like him. They think he's a wimp."

"Who took the pain?"

"Lisa is the one. She won't let them hurt him. She don't want to let anything happen to him. She came to help a long time ago. She was the first one who knew him."

"It's hard to take the pain for another person. What did you do for him?"

"I tried to take the pain for him. I tried to get him to move over but he couldn't always do it."

"What do you mean, make him move over?"

"Shirley, she made him scream. She tied him up and frightened him and shoved a dildo up his butt. There was blood. She tied him to the four corners of the bed on his stomach."

"Can you draw it?"

Terry drew a stick figure spread-eagled on a bed.

"She forced the dildo into him. Then she beat him on the butt with a kind of hammer to shove it in further."

"Why did she do that?"

Leo's voice had changed in a subtle manner. Still feminine, it now had a harder edge. "To cause pain and because she liked to watch him squirm. She was sick."

"How big was the dildo?"

"It was quite big."

"Could you draw it on this legal pad?"

"The paper is not long enough." He then drew an enormous ribbed dildo on a large manila envelope.

"What was it made of?"

"It was made of hard rubber. She would laugh. She thought it was funny."

"Was anyone with her?"

"No, she was always alone. She waited until everyone was gone and then Leo was scared. She said she'd cut his head off, and she tried to force him to cut off his own fingers. He wouldn't do it, then she'd hit him, lock him in the closet all night, wouldn't give him food or let him go to the bathroom."

"Why did she try to get him to cut off his fingers?"

"He had flipped off [indicating his middle finger on his right hand] his stepsister [Shirley's daughter] and this was punishment for it. She wanted to hurt him all the time. She took a big butcher knife and told him to cut off his own middle finger. She took scalding water and gave him enemas with it. He would scream. I tried to take all his pain, and I made him go away. It hurt. God, it hurt."

"Did anybody else know what was happening to Leo?"

"The school knew. Shirley had beaten his hands so badly with a ruler that they were black and blue and swollen. They hurt. He went to the school nurse to put them in cold water. She asked what happened. He said he fell down the stairs. He was afraid to tell. The nurse called Shirley and she said, 'He didn't fall down no stairs, I beat him.' So they knew, but they didn't do nothing."

"Did Leo ever have any bleeding from his penis?"

Leo nodded.

"Did anyone ever shove anything up there?"

He nodded. "Shirley used to shove a thermometer up his penis. She would tie him to a bed on his back and then she would play with him to give him an erection and then would shove a thermometer up when he was about ten years old and leave it there for ten to fifteen minutes to half an hour. She taped it in so it could not be expelled. This caused bleeding and infection. Sometimes he would pass what looked like meat."

In Leo's medical records, there had been visits to a doctor for blood clots coming out of his penis. Leo had been referred to a urologist who diagnosed a stricture (scarring) of the urethra, the tube that carries urine from the bladder to the tip of the penis. There was no mention of abuse in the medical records. No thought had been given to the questions of why a ten-year-old boy would have a stricture.

Dr. Lewis asked Leo to sign his name on the legal pad. He signed, "Lisa."

"Where do you stay?"

"I'm not far from Leo. I oversee all. David and Tony try to hurt Leo."

"The act for which he is in here—did he do it?"

"It was an accident. Tony and David didn't mean to hurt him but they got him into more trouble than they meant to. They didn't know how to handle the situation." Tears again began to stream down his cheeks silently.

"Lisa, could you go back into the void? There are several questions I would like to ask Leo." Leo suddenly opened his eyes, turned to us smiling, and the spell seemed broken. He spoke in a masculine manner; his smile suggested he had been relieved of a great weight. He seemed happy and content and was able to answer in a voice of lower pitch, without the feminine gentle, soft voice of Terry or the harsher feminine voice of Lisa.

Asked about his urological problems, he said that he had required a dilatation when he was ten. The medical records indicated that he had an anomaly of his upper urinary tract with a double ureter on one side and a constriction (scar) in his penis. He needed to have dilatation procedures because of scarring. This was done following a cystoscopy for passing red material (clots) when he was a child.

We showed him the picture of the dildo after he became Leo again and asked him what it was. He said he did not know and did not recognize it.

Leo's brain damage impaired his judgment as well as his ability to control his behaviors and appreciate their consequences. His bizarre thinking compounded this impairment. One personality was polite, affable, and intelligent. The angriest, the one who "took the pain," was dangerous. The nature and relentlessness of the abuse he had sustained as a child was the cause. Extraordinary ongoing physical and sexual abuse may be associated with the development of alternate personalities.

The host personality is usually quite benign—pleasantly charming and rational—very clearly not psychotic. Yet often the host hears voices that no one else can hear. But antipsychotic drugs like Haldol, Risperidol, and Thorazine—which very effectively suppress the auditory hallucinations of schizophrenia, mania, depression, and the toxic-metabolic encephalopaties that neurologists treat—do not resolve the auditory hallucinations of DID patients.

Symptoms such as blackouts, memory lapses, and staring spells that, given the history of brain injuries and possible seizures, seem as though they may indicate seizures, do not respond to anticonvulsant medication in DID patients. All these symptoms can represent dissociative episodes.

Auditory hallucinations and voices did not respond very well to antipsychotic medications in Leo's case, nor did his lapses respond to anticonvulsants. The multiple personalities of his DID that the voices represented would certainly explain his switch from confessing to pleading innocent and having "forgotten" the confession. Although he had not been found guilty of it, it would be easy to see why he might have chosen an older woman to rape and murder—in fact, such an act had been predicted by a psychologist who had seen Leo when he was younger, basing his opinion on the abuse Leo had suffered.

At one time Dr. Lewis and I believed that such memory lapses, which were described by many of the violent prisoners we had seen, represented a kind of complex partial seizure. It now seems much more likely that these are dissociative episodes and that dissociation with or without the full development of DID is common among violent prisoners.

Leo killed more or less in self-defense. The serial "cat burglaries" in which he was engaged provided for him the same kind of "high" that is experienced by many of the serial killers that we have seen, for many of the same reasons. The horrible abuse in childhood generates feelings of hopelessness, helplessness, victimization, and worthlessness. Brain damage leads to school failure and career rejection. Interpersonal deficiencies arising from dreadful handling in childhood cause the victim of such abuse to suffer from low self-esteem and to pursue power through other acts, often of a criminal nature and often having an idiosyncratic significance as, for example, the repeated theft of soiled women's undergarments. This seemed related, in Leo's case, to his enforced use of his abusive mother's soiled underwear for his sexual release at the end of their sessions.

Serial murderers often follow a particular ritual that has a specific symbolic meaning for each of them, originating in the sort of sexual and physical torture exemplified by Leo's childhood history. Why was Leo not a serial murderer? He was a serial "cat burglar" and probably did kill at least one elderly woman. There may have been other victims, about whom we have no information. Still, the question persists. Why might one person become a serial murderer and another something else? The truth is we do not know.

I have encountered a lot of skepticism regarding the existence of DID as well as the abuse that engenders it among my professional colleagues—particularly when the subject is a murderer suspected of malingering. Dr. Lewis and I, with colleagues, gathered data from medical, psychiatric, social service, school, military, and prison records; reviewed handwriting styles; and interviewed the subjects' family members and others.[4] This study corroborated signs and symptoms of DID in twelve murderers whom we had examined and documented severe abuse in eleven of them. The subjects had amnesia for the worst of their abuse and systematically underreported what they remembered. This sample of twelve exhaustively documented cases represents about 5 percent of all the murderers we have studied, but I would estimate that dissociation has occurred in many more, possibly a third of all the killers I have examined.

Only a minority of those who have dissociative episodes have evolved a full multiple personality disorder, but dissociation is a real phenomenon. Multiple personality is an extreme form of dissociation, a response to extreme abuse. When present in individuals with other mental illnesses and neurologic deficits, it can be very malignant indeed.

CHAPTER 11

Hitler and Hatred

Every corner of the globe has torturers in some position of authority: the police who beat and maimed a Haitian immigrant in New York City; the former testicle-crushing dictators of Central and South America; Milosevic and his tyrannical rule in Serbia; Stalin; Saddam Hussein—the list goes on and on. From where does this hatred and violence come? Is it possible that the violence that is expressed politically, ethnically, racially, and religiously shares common determinants with the violence committed in an urban parking lot? I think it is likely that ostensibly political and criminal violence spring from a similar source.

The most uncontrolled and far-reaching expression of hatred in the twentieth century has been that of Nazi Germany and the Holocaust. I have always wondered what sort of person could perpetrate crimes against innocent men, women, and children in concentration camps: poison gas, burning, starvation, death marches, unrelenting forced labor, beating, sexual abuse, lethal medical "experiments." Who could order such things? Who could carry out such orders, or even torture these victims autonomously?

Goldhagen, in his book *Hitler's Willing Executioners*,[1] emphasizes the large numbers of ordinary Germans who actually participated in the Holocaust, by killing, torturing, and maiming Jews. He estimates that as many as 500,000 Germans were active perpetrators of murder. This is a huge number, but in a country of sixty million why did some participate in the killing while the majority did not? One possible answer might be that many of these killers could have been raised in hatred-filled and abusive households. Anti-Semitism may have been imbued with mother's milk, along with wife and child beating and other intrafamilial violence. I envision members of the death squads as young men who had been raised in lower-middle-class homes by brutish parents and whose fathers returned from work drunk, beat their wives and children, fell into bed in a stupor, and repeated this sequence the next day. I would speculate that the murderers had themselves been abused, and their willingness to murder derived from this and the anti-Semitism of their environment. Societal approval of anti-Semitism and governmental approval of murder lifted the lid on the abuse-engendered impulse of the offenders. This theory has grown out of my research. It explains why the ordinary Germans who were not neurologically or mentally impaired carried out the urge to murder Jews under the Nazis. Many perpetrators, such as brutal concentration camp guards, could also have had neurologic or mental illnesses that augmented the viciousness of their behavior once the societal lid that controls murder had been removed by the government so far as the Jews were concerned.

Frank McCourt's book *Angela's Ashes* offers insight into how abuse might lead to bigotry.[2] The author movingly portrays the poverty into which his father's alcoholism and his mother's depression had thrust the family. His father would return home late at night, intoxicated, his paycheck gone. He would awaken his starving children and have them stand in the kitchen and recite noble poems and sing patriotic songs that celebrate the Irish and condemn the English as the cause of the misery of the Irish and, by extension, of his family. As a reward, the father gave each child a penny with which to purchase candy the following day.

Their father successfully displaced his own responsibility for the family's poverty to the English. Fortunately for the author, his father and mother were not violent and abusive, but what if they had been? Could deprivation and abuse be the origin of IRA terrorism in lower middle-class Belfast?

The dreadful sense of helplessness and humiliation that is engendered by child abuse, the victim's sense of powerlessness and fear, and the rage that spring from it are crucially important motivators toward violence. Depression in an abused person intensifies this dynamic. The brain damage and/or intoxication that can be superimposed interferes with the capacity of the abused individual to control the expression of his or her rage and hatred. The paranoia and delusional thinking of individuals who have additionally inherited mental instability and mental illness exacerbate these dark feelings and abolish the capacity to love, to trust, and to enjoy life. So fundamentally do these factors harm the psyche that the life's work of such victims can be seen as an attempt to escape from their victimhood and sometimes "rise" to the level of perpetrators.

Some of the murderers I have evaluated in prison have identified themselves as white supremacists and neo-Nazis. There is a large fellowship of such men in prisons. Blacks and Hispanics have parallel organizations in prisons that direct hatred at other groups. One condemned man's misogynism was of such intensity that it reached the level of potential violence shared by Nazis and racists.

Trent Scaggs surprised me when he asked, "Why was I born? What is the point of my life?" I was unprepared for these questions. Trent was a thin, pale Caucasian male, only 5 feet, 4 inches, with the facial stigmata of fetal alcohol effect. When I examined him at his attorney's request, he was on death row for killing another inmate. Originally, he had been incarcerated for assaulting two of his female teachers. The first assault occurred when he was fourteen years old and in residential treatment, the second while he was in "reform" school, being punished for the first.

He had entered the small examination room with shackles on his feet, his hands immobilized by handcuffs attached to a chain around his midsection. We were alone in the barren room. The only furnishings were two chairs and a table that were all bolted to the floor. He had been accompanied by a female guard who refused to uncuff his hands; she said that when I was ready to examine his hands another guard would remove the cuffs and remain in the room.

I had not answered his questions, not because I did not wish to say anything but because I could think of nothing to say. Perhaps he expected no answer, because he continued: "The women here, the guards, are not to be trusted. They drive me up the wall. With two Uzis I would kill them all. I've been dominated all my life. I can't get a grip on myself or get people to respect me or leave me alone."[3] This was an unusual introduction. I followed the conversational lead.

"Do you feel the same way about the male guards?" I asked.

"No," he replied with barely suppressed emotion. "I have a hatred for women. They are responsible for what happens in the prison, the way they walk and talk. They are sneaky. They talk about you behind your back. They tease. They put you on suicide watch for a week in a cold, empty cell, stark naked. It is not to keep you from committing suicide but for punishment." With a grim facial expression he continued, "The psychologist, Peggy, has you put in suicide watch just for leaving an apple in your footlocker. . . . I'd like to drown her in a toilet bowl."

There was no smile accompanying this fantasy. It was literal, a means of causing Peggy to die in a humiliating manner.

"They think I am weak because I am gay," Trent said. He denied this angrily. "I am not gay! I have to do what I do in order to get along. I am not strong. I can't protect myself from these big guys. I tried to build up my muscles by lifting weights, but it doesn't work. It doesn't help . . . I don't like what I have to do. I hate it. I'm all tore up inside. I've been ruined. My bowels don't work. I can't take a shit more than once a week. It hurts me to shit."

I asked if he had a protector in prison. Trent said he did, but his "protector" gave Trent to other prisoners in return for favors Trent was his sex slave. . . . "It's not right to make X-rated movies," he said. "Pornography is caused by women. They cause the whole thing in here. The other prisoners look at these sex magazines and then they make me put up my legs in the same way as the women they see in the magazines, and they fuck me in my ass or make me suck them off. I hate it, but it's the only way I can stay alive in here."

I asked if he remembered killing the other prisoner for which he was on death row. He spoke with a puzzled facial expression as he said, "Yeah, he was a little guy just like me. We were having sex with each other. He had already fucked me from behind and I was doing it to him when I suddenly 'lost it.'"

"But why?" I asked. He had no explanation. He hung his head and said quietly, "Maybe it was for control. I couldn't stop myself. I grabbed Billy's head in a chokehold from behind and strangled him to death. I had nothing against him. He was just a little guy, like me."

Although this was the only homicide Trent had committed, there were many other violent offenses in his record. All had been assaults on women. He fantasized and dreamed about beating, shooting, killing, drowning, and torturing women. His hatred of women—all women—and his paranoia about their actions, their presumed negative attitude toward him, and their responsibility for the horrible conditions of his life were reminiscent of political murderers and their misdirected hatred of groups of different religion, race, and ethnicity.

His hatred of women was as irrationally directed as well. Despite his focus on women, his suffering was, in reality, mainly the result of the actions of men. At the age of three, Trent and his brother had been removed from home by his maternal aunt, by court order, because his biological parents, were abusive alcoholics. The precipitant of this change was his father's heating a knife on the open flame of the stove and burning Trent's buttocks as a punishment. It left a permanent and verifiable scar. After this, the family, such as it was, and the court agreed that a change in environment was advisable.

Unfortunately, Amanda's husband, Lyle, was also an alcoholic who beat Trent and his brother savagely and brutalized them repeatedly, using a belt, his fists, and other instruments. Once his uncle used of a monkey wrench to hit Trent in the head; the blow was so forceful that it left a prominent scar that was visible and palpable when I examined him at the time of my visit three decades later.

When his uncle grew tired of beating him, he forced Trent to stand on tiptoes in the corner of a room, naked. Trent was not allowed to turn around or to let his heels touch the floor. If either occurred, Uncle Lyle walloped Trent with a belt. Because Trent could not turn around, he did not know when his uncle was observing him to monitor his compliance. The blows came without warning and these torture sessions lasted for hours at a time.

One of the most destructive of his uncle's abuses was to force Trent and his brother to fight with each other, viciously, like gladiators, for his entertainment. Trent's aunt was powerless to interfere as she was subjected to beatings as well.

Without wishing to diminish the horror of the beatings, I considered the very worst aspect of Trent's torment at the hands of his uncle to be repeated sodomy. Lyle forced him to perform oral sex and to submit to anal sex when Trent was only four years old.

The offense that first brought Trent's plight to the attention of the authorities occurred when Trent's brother had misplaced Lyle's cigarettes. Lyle mistakenly blamed Trent and told Trent to place the fingers of his left hand on the kitchen table, leaving the thumb and the palm free. Not knowing what was coming, Trent complied. Lyle then raised his right hand that held a concealed meat cleaver and attempted to strike Trent's hand and sever all four fingers. Fortunately, Trent flinched at the last second and withdrew his hand. The meat cleaver missed three fingers but nearly severed Trent's index finger at the base. Amanda brought Trent to the hospital to have his index finger sewed back. This operation was successful, but the investigation of the circumstances of the injury led to Trent's removal from the home of Lyle and Amanda.

Trent was then housed in a variety of institutions. He was extremely hyperactive, unresponsive to discipline, and destructive. He was severely punished and abused for his bad behavior in some of the foster homes in which he was placed. The only other institutions available for him were mental institutions and reform schools. He was raised by the state but no placement in the many institutions and homes in which he was domiciled was successful in altering his behavior.

A number of psychological and physical assessments revealed borderline retardation, learning difficulties, and probable fetal alcohol effect. The latter caused small stature, microcephaly, intellectual inadequacy, and facial abnormalities typical of that condition. The extraordinary abuse by his uncle, beatings in foster homes, and rapes in the institutional settings were all documented in the voluminous records that had been collected prior to and after his first violent crime.

On weekends and vacations over the years until he was fourteen, Lyle would remove Trent from the institution in which he was housed for short visits "home." Trent described to me his terror at the onset of such visits. He tried to sleep in a bed near a window so that he could look out for the approach of his uncle and try to hide himself to prevent his release. How the authorities could allow this when they had removed Trent from Lyle's care for such extremities of mistreatment, I do not know, but it happened.

The purpose of his uncle's attentiveness was not to provide Trent with a pleasant weekend at home, a break from the rigors of the reform school, but rather to subject him freely to perverted sexual abuse in Lyle's home. These activities included sodomy, oral sex, and the attempt to recruit Trent in assisting with the forcible rape of his aunt, Amanda.

The excuse provided by Lyle to Trent for sodomizing, raping, and assaulting him was that Amanda did not provide Lyle, her husband, with an outlet for the sexual urges and desires that she purposely provoked in him. If she had behaved sexually as she was supposed to, according to Lyle, Trent would have been spared all torment at Lyle's hands.

This made Trent very angry at his aunt. He joined Lyle in his hatred of Amanda and women in general. He adopted Lyle's view that women sexually taunted men.

Indeed, he reported that Amanda did behave inappropriately. According to Trent, she wore a diaphanous negligee and let Trent see her in other states of partial undress. She often invited Trent into the bathroom while she bathed so he could wash her back, and she bathed and soaped him during his baths. Though this was the closest Trent ever came to experiencing gentleness and affection, he hated his aunt's attentions and he hated her. During one of her bathing sessions, he planned to stab her with a knife, but his uncle's unexpected arrival prevented Trent from executing his plan. But Trent's hatred of Amanda became the paradigm for his hatred of all women.

Because Trent was small of stature and slight of build, he was a target for homosexual rape in the institutions in which he was placed. He ran away several times to escape this and the beatings by other, bigger boys. Each escape attempt resulted in his recapture, and he would be punished by being held in more secure, state-owned institutions.

Trent said that during one of his escapes from reform school, he met a girl named Mary who was also "on the run" from an abusive home. She was his only true love, the only woman he ever loved or could love. He fantasized about her and wanted desperately to know what had happened to her and to see her, but despite repeated attempts by investigators for his defense to trace her, she was never found. Perhaps she was a real person, perhaps a fantasy of his.

After he was recaptured, Trent said he received an urgent telephone call from Mary, begging him to protect her. There is no record of such a call, but Trent did attempt another very poorly planned escape.

He climbed out of the room he had been assigned, shoeless to avoid making noise, crossed a metal parapet, and entered an attached schoolroom in which he planned to hide before leaving the institution later that night. How he was going to run away without shoes was beyond the scope of his ability to plan.

The authorities quickly noticed his absence from his quarters and launched a major effort to find him. His female teacher unwittingly entered the classroom in which he had secreted himself. He told me that she was very happy to have found him. She gave him a big hug and said that everyone had missed him, was worried about him, and was looking for him.

He had a sudden feeling of empowerment. He roughly grabbed

her breasts and tore her clothes. She screamed; to quiet her, he choked her and nearly killed her. He told me that he had been accused of attempted rape but, in denying this, he said, somewhat affectingly, "I didn't know what else to do except for grabbing her. I did not know about sex, only what my uncle had done to me." Others, alerted by her cries, came to the rescue. He was apprehended, tried, and punished with incarceration in another institution.

When Trent later attacked another female teacher, he was incarcerated in a real prison at age sixteen. There he attacked a female guard, Ginny, who had ridiculed him, calling him "Snow White" and branding him as a homosexual. He wanted to rape and humiliate her in revenge for the degradation he felt by her remark. He said, "She would never be able to live it down if I raped her."

He had the opportunity to attack her when they were alone in his cell on one occasion. He beat her, cut her forehead, and threw her to the floor. With satisfaction he told me, "She wet herself with fear of me, and she was bloodied." He kept hitting her, "to make women pay" and to let her know how he felt. He was "really getting into it," as this was the way "to get back at the male guards too, to beat their women." He was sentenced to fifteen extra years in prison for this and, while serving the time, he murdered his fellow inmate, Billy.

Trent did not relate all the details of his attacks on others; much of what I knew was derived from reports. He may have been in a dissociative state during the attacks, or maybe he did not wish to share with me his feelings at the time. Clearly, he lost control of himself with Billy and he had no reason for the attacks on the teachers except his own internal pathology. His victims had provided no offense. Even though he did have some reason for hating the guard, Ginny, I believe that all his attacks were motivated by his hatred of women and his desire to exercise violent control over them. Even the killing of his fellow inmate was a kind of murder of a woman.

My examination revealed mild mental slowness, poor motor

coordination, many evidences of frontal lobe dysfunction, the stigmata of the fetal alcohol effect, and many scars that confirmed the history of abuse that he recounted and that the records described or implied. Psychological tests confirmed his borderline mental retardation. His IQ was 69, and there was frontal-executive dysfunction. His MRI was abnormal, showing white matter abnormalities in both frontal lobes.

What purpose was there to his life? That was the question he posed to me at our introduction. He could not identify any positive function his life provided to him or to society.

What if Trent had heard a political leader say "Women are our misfortune!" just as Hitler said "The Jews are our misfortune!" A public condemnation of women and homosexuals as the hereditary carriers of social pathology would have dignified Trent's suffering and provided a social outlet for his hatred.

What if a Hitler-like political leader said that women make men work and then steal the wealth that men amass, that they create pornography, transmit infectious diseases, are weak, subhuman, and defective? Hitler said this of the Jews, gypsies, and homosexuals. Such a message would probably have been as welcome to Trent as Hitler's was to many Germans.

What if Trent were released from prison and told that he had a glorious role to play in saving civilization and in creating a new and just society? That he, miserable Trent Scaggs, would be a leader in the elimination of women and homosexuals? Imagine how Trent might behave if he were put in command of a camp in which these "subhuman species" were concentrated, where he would be able to beat, rape, torture, maim, and murder as he wished, his actions being condoned as ridding society and the world from the problems caused by females and homosexuals.

Is there some common element between the case of Trent Scaggs and the rise of the Nazis under Hitler? I have little data with which to answer that question, but the insights I have gained from my work with criminals suggests that Adolf Hitler and Trent Scaggs had a lot in common.

In the early 1940s, the Office of Strategic Services (OSS), the precursor of the Central Intelligence Agency (CIA), asked psychiatrist Walter C. Langer to provide a psychological analysis of Adolf Hitler on the basis of information they had available. Dr. Langer cooperated and prepared a report in early 1944. For many years unknown to scholars and the public alike because of its "secret" classification, even after the end of World War II, the report was published in 1972.[4] The study is both impressive and fascinating. The data amassed by Dr. Langer indicate that Hitler was abused as a child, very paranoid, and probably manic-depressive.

Adolf Hitler's father, Alois, was generally described as a very domineering individual, virulently anti-Semitic, a veritable tyrant in his home who used to beat the children unmercifully. It is alleged that he once beat his older son (Alois Jr.) into a state of unconsciousness and on another occasion beat Adolf so severely that he left him for dead. He was also considered to be a drunkard, whom the children frequently would have to bring home from the taverns. Once home, a huge scene would take place, during which he would beat wife, children, and dog rather indiscriminately (Langer, p. 104).

In *Mein Kampf*,[5] Adolf Hitler describes scenes of a child's life in a lower-class German family marked by alcoholism, physical abuse, and hints of sexual abuse:

> Among the five children (there is) a boy of, let us say, three. . . . [The] battle is fought between the parents, and this is almost daily, in ways whose inward coarsesnes is extreme, the results of such an object lesson are bound to appear in the children. . . . [W]hat must the results be if the dispute takes the form of father's brutality to mother, of drunken maltreatment, a person who does not know the life can hardly imagine. By the time he is six, the pitiable little boy has a notion of things which must horrify even an adult. (p. 44)

The child in this passage is probably Hitler himself. Indeed his description becomes more specific and personal:

> Things end badly if the man goes his way . . . and the wife opposes him for the children's sake. Then there are quarrels and bad

> blood, and the more the husband drifts apart from his wife, the nearer he drifts to alcohol. Every Saturday he begins to be drunk; and in self-preservation for herself and the children the wife fights for the few pence she can snatch from him, and those are mostly what she can get on his way from factory to saloon. When at last he comes home himself on Sunday or Monday night, drunk and brutal, but always relieved of his last penny, there are likely to be scenes that would wring tears from a stone. I saw all this go on. (p. 41)

Hitler, describing his own childhood behavior, provided hints of conduct disorder and oppositional defiant disorder:

> Constant romping around outdoors, the long road to school, and an association with extremely robust boys, which sometimes gravely worried my mother, all combined to make me anything but a stay-at-home. . . . I believe that even then my oratorical gift was being schooled by more or less violent disputes with my playmates. I had become a little ringleader who . . . was fairly hard to handle. (p. 21)

But Hitler's violence was likely engendered by more than abuse alone. Like Trent Scaggs, who was paranoid and neurologically damaged, there is evidence that Hitler suffered from mental illness with paranoia. He had tremendous mood swings and was said to fly into rages and tirades on slight provocation. A contradiction, a criticism, or a doubt concerning the truth or wisdom of something he had said or done, the anticipation of opposition, or a challenge only by implication might trigger an uncontrolled display of anger.

The descriptions of his behavior during these rages are vivid. He would bang his fists on the tables and walls, scolding, shouting, and stammering; on some occasions, foaming saliva gathered in the corners of his mouth. His hair became disheveled, his eyes fixed, his face distorted and purple. In extreme cases he was said to have rolled on the floor and chewed on the carpets. But William Shirer, in *Berlin Diary* reported that in 1938 Hitler did this so often that his associates referred to him as *teppischfresser* (carpet eater).[6] Even without this added detail, his behavior was still unmodulated.

Other facts bolster the theory that Hitler suffered from a mood disorder. Although he could have supported himself by painting postcards, he lived as a penniless vagrant for almost seven years, before joining the army after the outbreak of World War I.

This suggests a prolonged period of depression. On the other hand, his rages and grandiosity and irritability are all consistent with a diagnosis of mania. Mania may have also driven him to work for several days on end with little or no sleep. Indeed, Langer provided evidence of a pervasive sleep problem in Hitler that is common in affective disorders. His habit of taking a sedative before retiring and, during the war, using amphetamines to stay alert adds further credence to this theory.

Hitler's paranoia was famous as well—with relation to Jews primarily, but also regarding homosexuals, gypsies, and the retarded and psychiatrically ill. He was described as haunted by secret misgivings and distrusted everyone, even those closest to him. He could not establish any close friendships for fear of being betrayed. "I am the loneliest man on earth," he told an employee of his household.

There is no evidence of brain damage in Hitler's youth. Langer said that Hitler's early report cards showed an almost unbroken line of A's in all his school subjects. However, at the age of eleven, he suddenly failed most of his classes and had to repeat the year. This may have related to mental illness—the second decade of life is commonly the age of onset of bipolar affective disorder and schizophrenia. So even without neurologic deficits, it is likely that abuse, paranoia, and mental illness were all factors in the formation of Hitler's personality and character.

According to Langer, Hitler's paranoid anti-Semitism targeted a convenient and generally unpopular outlet for his aggression and desire for domination. He could be a perpetrator rather than a victim. He projected all the evil he had experienced and which he hated onto the Jews, whom he believed to be blighted genetically, infectious, and the cause of all his pain and the ills of society. He may have acquired his attitude in his home, and the intensity of his feeling could have been the result of abuse, a combination of an unhealthy childhood marred by abuse.

The early life experiences of many individuals involved in hate groups and terrorism may very well be similar to those of Trent Scaggs and Hitler. For instance, according to Mustafa Masri, a psychiatrist in the Gaza Strip who plans, supervises, and evaluates clinical work at the Gaza Community Mental Health Program,[7] "Domestic violence is endemic in Gaza. Men have a lot of aggression and very few channels for ventilating it. . . . Men—husbands, sons, fathers—use women to ventilate and beating is part of the culture." Said Dr. Eyad Sarraj, head of the program, "Ours is a harsh, underdeveloped culture. It is harsh in the way it treats its members. There is widespread abuse and violence. The suffering is not limited to the individual (with posttraumatic stress disorder) but is often passed on through violent behavior directed at the women and children in the household."

Is it not possible that there is a relationship between violence in the home and terrorism? Those in any society who support or participate in murder and terrorist acts have the "respectability" of being violent for political or religious reasons, but the triad (child abuse, brain injury, and paranoia) does not know political, ethnic, or religious boundaries and could find its outlet in a "worthy" cause.

In a society that permits or encourages the stigmatization of a group of people, mental illness and or brain damage may not be necessary to produce homicidal violence. The violent impulses generated by abuse could be expressed by medically normal abused individuals in a sick society.

The unrestrained approval of violence in certain political parties and gangs may make such groups attractive to the abused. Although we have very little information about the family dynamics of the members of terrorist organizations, I believe that the history of physical and sexual abuse, and even mental illness, paranoia, and brain damage, is prevalent among them. If a society (defined as a country, political party, or a group) removed the demand that its members behave in a nonviolent manner toward its enemies, brain damage and mental illness would not be necessary to unleash the impulses to violence that were engendered by early abuse. If society says, "You are now free to rape, torture, and kill our enemies," the abused may respond with enthusiasm. Not every American soldier in Vietnam committed atrocities, nor did every German in World War II, nor every Arab in the PLO. What is the difference between the violent and nonviolent members of such societies? I would postulate that early childhood abuse distinguishes perpetrators from nonperpetrators.

In a sociological study of members of racist organizations in the United States, obtained at no little personal risk,[8] Raphael Ezekiel pointed out a number of features in the backgrounds of members of hate groups that are reminiscent of the life of the very violent offenders I have evaluated. Most of Ezekiel's subjects had lost a parent by age seven, usually the father. Stepfathers and the mother's boyfriends were typically described as cold, rough, and abusive. Connections with family outside the nuclear group were frail, and there was little parenting in depth through an extended family. Church and social agencies (other than police) played little or no modulating role in the families, and ties between siblings were minimal. Schooling was interrupted, with early truancy and dropouts prevalent.

Ezekiel wrote, "A few members spoke spontaneously of paternal alcoholism and violence . . . I think that a lot was held back from me. [Some] members reported alcoholism, . . . family violence, [and] serving time at detention centers, jails, or prisons. I am fairly sure . . . that there was a good deal more penal time than this. Several had been farmed out to foster homes during childhood." Alcoholism, violence, mental illness, paranoia, school failure, and medical/neurologic problems run through Ezekiel's entire account, suggestively reminiscent of the factors I have encountered in violent offenders.

The concept that violent impulses are engendered by the experience of abuse and allowed expression in those abused individuals who also have endogenous neurologic and psychiatric vulnerabilities has very important implications for the reduction or prevention of violence. Efforts directed at mitigating any one of the interacting factors that cause violence would be likely to have a major impact. The most effective action and possibly the easiest to accomplish, though, would be the prevention of child abuse through social and educational programs. A major step in the primary prevention of violent crime can probably be achieved by a national effort to reduce or eliminate all use of corporal punishment by parents and teachers.[9]

CHAPTER 12

Prevention and Treatment

I preside over morning report, a teaching-working exercise attended by medical students and neurology residents, during which residents present cases of the patients examined in consultation during the preceding day. On this particular morning, the resident who had been on duty the night before told us that the police had brought a man to the emergency room (ER) and the ER staff had requested a neurological consultation to rule out epileptic seizures.

The man had been lurking around a Georgetown store, the Gap, ostensibly examining merchandise in the women's section that evening among other shoppers, mostly women. Approaching a woman (to whom he was a complete stranger) from behind, he suddenly placed both his hands on her buttocks and then barked several times aloud. His face and body twitched dramatically. The woman and others in the store were startled; the woman screamed; the manager called over a police officer who had been standing just outside the store; and the offender was arrested. The twitching and vocalizations of the man were so extreme that the policeman thought he might be an epileptic, so the prisoner, now a patient, was taken by ambulance to our hospital.

This story of a man with obvious Tourette's syndrome evoked smiles among the students and residents who heard it, though all were quite sympathetic to the startled woman who had been assaulted. The man had entered the store gripped by a recurrent, obsessive thought about touching a woman's buttocks. He had wandered in the store trying to suppress his desire but finally he lost control and was compelled to carry out the act. The vocal ejaculations and twitches were the motor manifestations of Tourette's, and the obsessive-compulsive symptoms were the cognitive-behavioral manifestations of the same illness.

The condition is familial and its symptoms are treatable—the motor symptoms with haloperidol (Haldol), the obsessive-compulsive disorder with fluoxetine (Prozac). Medically, the case was not especially challenging, but I was taken aback when a student asked, "Is the patient responsible for what he did? Should he be punished or treated?" In my hurry to get on with the day, I simply replied, "Both." My response ended the discussion but the question remained. It is a very good question and goes to the heart of the apparent incompatibility between the medical-scientific approach and the moral-religious-legal approach to understanding violence.

Doubtless many men have wished to touch the buttocks of a strange woman. But people who enjoy touching or looking at the sexually stimulating portions of other's bodies can control their urges without too much effort. However, because their brains are probably constructed differently, the urges of people with Tourette's syndrome are more intense and much more difficult to control. Obviously, not all people with Tourette's sexually assault strangers and those who wish to can usually control those urges. But if this patient with Tourette's had sustained brain damage or suffers from a mental illness in addition to Tourette's, it might be even more difficult for him to bottle up or divert his urges. Yet by treating Tourette's or other concomitant brain disorders, we might lower the chances of his recidivism. But how responsible is he?

No treatment program directed at violent persons has ever been proven to eradicate their potential for violence, and it is difficult to believe that the rehabilitation of most of the murderers whom I have seen would be possible. Most neurologically damaged individuals, most abused individuals, and most individuals prone to violence have not been turned into automatons but retain free will of considerable scope. To the extent that they have exercised their free will inappropriately, they should be held responsible and punished proportionately. Even if the death penalty is inappropriate for such cases, society must be protected from them.

Mitigation of the death penalty to life imprisonment with medical supervision is not equivalent to "getting off" and going free. Some say prisons are like country clubs, saying "They have TV"—a comment about prison life that makes it seem quite plush behind bars. I have visited dozens of prisons in the United States. None are remotely like any country club I would join. The loss of freedom is a terrible punishment, no matter how humane the conditions of confinement, and most prisons do not provide benign environments. The food, the company, the rooms, and the loss of control over basic daily life (when to bathe, eat, sleep, watch TV) are all horrible. There is constant danger of physical and sexual assault. Infectious diseases, like AIDS and tuberculosis, are common.

The medical-scientific approach to homicide should not be a substitute for the law. Nor will it always be mitigating or exculpatory, because punishment would depend on the efficacy of therapy. Most issues of guilt and punishment can and should be considered in the light of medical-scientific information and modified accordingly.

Treatment and punishment are responses to crime that certainly have less potential for eliminating violence than programs designed to prevent it. Such prevention programs can follow the pioneering trail of Hawaii's Healthy Start program, founded in July 1985 as a demonstration project designed to prevent child abuse and neglect in Oahu.[1] The program served a multiethnic community with mixed urban and rural population in an economically depressed area characterized by substandard housing, underemployed adults, substance abuse, child abuse, and neglect. Three years later not a single case of abuse and only four cases of neglect among the project's 241 high-risk families had been reported. By July 1990, Healthy Start services were expanded to eleven sites by the state legislature to reach approximately 52 percent of at-risk families of newborns throughout Hawaii.

The Healthy Start approach was designed to prevent child

abuse and neglect by improving the mother's coping skills and functioning, promoting her parenting skills, and facilitating her ability to interact with her child. Several complementary features constituted this approach, including systematic hospital-based screening to identify high-risk families of newborns, contacting at-risk families and encouraging them to accept home-visiting services, additional assessment interviews, supportive services focusing on problems discussed during the interview, and providing information and advice about the baby's care and development. During the three-year demonstration period, 95 percent of families voluntarily accepted the offer of services.

***Risk Indicators Used in Early Identification*[2]**

1. Marital status: single, separated, divorced
2. Partner unemployed
3. Inadequate income (per patient) or no information regarding source of income
4. Unstable housing
5. No phone
6. Education under twelve years
7. Inadequate emergency contacts (e.g., no immediate family contacts)
8. History of substance abuse
9. Late (after twelve weeks) or no prenatal care
10. History of abortions
11. History of psychiatric care
12. Abortion unsuccessfully sought or attempted
13. Relinquishment for adoption sought or attempted
14. Marital or family problems
15. History of depression

Once a family accepted the offer of service, a family support worker contacted the mother to establish rapport and schedule a home visit. Initial visits were usually devoted to building trust, assessing family needs, and providing help with immediate problems such as obtaining emergency food supplies, completing applications for public housing, or resolving crises in family relationships. Workers focused primarily on providing emotional support to parents and modeling effective skills in coping with everyday problems. Their "parent the parent" strategy allowed initial dependence before encouraging independence. Workers also modeled parent-child interaction.

The intensity of service was individualized based on the family's need and level of risk. All families received weekly home visits. The decision to change a family's level of need was based on criteria such as the frequency of family crises, the quality of parent-child interaction, and the family's ability to use other community resources. As families became more stable, responsive to children's needs, and autonomous, the frequency of home visits diminished. A family's progress to the least needy level resulted in quarterly visits that continued until the child was five years old. In a previous service program that stopped following families once they were no longer considered "high risk," cases of child abuse and neglect were reported later in a number of these families. Family situations can deteriorate, and the birth of subsequent children can add to family stress. Learning from earlier experience, the Healthy Start program was designed to maintain follow-up until the targeted child reached age five and entered school. At that point, the education system provided at least some link between the family and the larger community. Thus service intensity was constantly adjusted to the needs of the family, assuring that families who were doing well moved along, and those who needed more support were not promoted arbitrarily.

As its name suggests, the Healthy Start program emphasized preventive health care as an important aspect of promoting positive child development. Each family was assisted in selecting a primary care provider, which might be a pediatrician, family physician, or public health nursing clinic. Of the 241 high-risk families, 176 had received services for at least one year during the three-year demonstration.

This program seems to have been a dramatic success, but the results of this study should be scrutinized carefully before accepting the Healthy Start program as a national paradigm to reduce violent crime. This preliminary effort cost $2,500 per year per family and appeared to provide a promising avenue toward the social goal of preventing abuse, but there was no control group (i.e., a matched sample of similarly at-risk families that did not receive the services offered to the study group) to prove the effectiveness of the intervention.

The requirement of a control group may seem nitpicking or even unethical to some, but the history of medicine is replete with stories of expensive interventions that ultimately proved to be ineffective. Since so many abusers are boyfriends of the mother or stepfathers of the abused children,[3] it is difficult to understand how a trained nonprofessional working only with the mother could effect a safe and nonabusive home environment.

Other studies have bolstered the idea that preventive approaches are effective. These have been ably summarized by Elliott Currie in his book *Crime and Punishment in America*. Noteworthy preventive programs include the Prenatal–Early Infancy Program in Elmira, New York, started in the late 1970s by David Olds at the University of Rochester, and another program sponsored by the Johns Hopkins University Medical School serving low-income African-American mothers in Baltimore in the early 1980s. The promising but tentative findings of these studies have been confirmed by the National Committee to Prevent Child Abuse.

Most of the studies performed to date have not included long-term follow-up nor have they focused on the very highest-risk, inner-city families. Both Hawaii and Elmira—a rural, mostly white area—have relatively low rates of violent crime and poverty. Although abusive homes, brain damage, and mental illness—not poverty, the urban environment, and race—are the real causes of violence the triad is more common in poor urban areas.

While preventive action programs in Hawaii, Elmira, and Baltimore seemed to lower the rate of child abuse, it is impossible to know whether that translated into lower rates of violent crime in the community afterward. To determine whether a program directed at preventing child abuse effectively reduced subsequent violent crime, the program would first have to target high-risk families in high-crime areas, deliver services to some but not others over a period of years—from birth to five years at least—and have a follow-up of at least a decade, possibly two. A well-designed study with control groups might provide enough solid data to justify the widespread application of its methodology to at-risk populations in high-crime areas around the country. Some controlled studies have already pointed that way.

The Perry Preschool Program began in the 1960s and was directed to a low-income African-American neighborhood in Ypsilanti, Michigan.[4] The intervention was very modest. Poor children, aged three and four, enrolled in preschool for two and a half hours a day and teachers visited them at home once a week for an hour and a half. Most children were in the program for two years, others just for one. The design and evaluation were very carefully crafted: 123 neighborhood children were randomly assigned either to the program group or to a control group that did not attend the preschool program, and persistent efforts to keep up with both groups extended over two decades. The youngest of the children are about thirty years old now. The program graduates are more likely to be literate, employed, and not receiving welfare, one-fifth as likely to have been arrested five or more times, and one-fourth as likely to have been arrested for drug-related crimes.

Other groups have applied similar techniques to high-risk populations. These include the Yale Child Welfare Research Program and the Syracuse University Family Development Reseach Program.[5] The early follow-up evaluations of these programs were less encouraging than the late follow-up evaluations conducted ten years later, which revealed a reduction in rates of delinquency. Twenty-two percent of the controls and only 6 percent of the study children in the Syracuse program had a juvenile record. And the crimes committed by the study group were minor while those of the control group included robbery, assault, and sexual assault.

Not all the experiences with preventive programs have been as positive. The Beethoven Project set up in the very worst part of Chicago's South Side did not have any measurable impact on the later criminality or drug involvement of its participants. Nonetheless, the positive results of the Yale, Syracuse, Baltimore, Elmira, and Hawaii studies are encouraging.

Much of what we define as child abuse seems to be the result of simple ignorance. For example, some prison authorities have discovered that incarcerated mothers of infants do not know that the infant must be held during bottle-feeding, demonstrating that a skill one might have expected to be instinctual is actually learned. Many teenage mothers have never seen proper care delivered to an infant or child and do not know what constitutes neglect and abuse.

But corrective home visits and preschool experiences are considered to be unduly expensive frills in this country. In Scandinavia, France, and most of the developed countries of Europe and Asia, home visiting and preschool programs are popular and well supported by the government. Perhaps if this investment in the brain development of disadvantaged youth were made in this country, there would be less need for the penal beds that now cost about $30,000 per person annually. There are almost two million individuals behind bars in the United States at any given time.

Ignorant families often respond to the cries of their babies and toddlers with physical punishment. If the only technique for altering the behavior of others that the parent has seen or experienced involves violence, or the threat of beatings, then those are the techniques the parent will use, even if the punishment necessary to extinguish the child's annoying behavior is excessive or potentially fatal. Holding, feeding, changing, bathing, singing, rocking, playing, distracting, reading, talking, and carrying are potential parenting techniques that can and should be taught, perhaps even to teenagers in school as well as at-risk mothers at home.

It is not known how long a child can sustain abuse and still escape the deep psychological scars that lead to violence. Once an individual has become behaviorally disturbed, it may be too late to intervene effectively. Yet data from some studies indicate that intervention even at this time may prove effective. Removal of a child from the abusive environment, placement in a supportive and caring home, and treatment with medication for such cerebral diseases as depression, mania, and epilepsy all can have positive results.

My own experience suggests that unusual destructiveness of children toward property, aggressiveness toward others with fighting, the torture of animals, arson, enuresis, and encopresis are the hallmarks of the previolent. Children with conduct disorder and oppositional defiant disorder should be the subjects of careful medical and social investigation and prompt, effective intervention if abuse is uncovered.

The recognition and treatment of mental illness in children must have a high priority, refuting the prejudice that all psychiatric symptoms in children are just phases of behavior within the spectrum of normality. The family history of serious mental illness furnishes an important diagnostic clue to the source of a persistent behavior disorder. Schizophrenia, mania, depression, obsessive-compulsive disorder, brain damage, and developmental delay all produce behavioral and cognitive symptoms and sleep disorders, and all can be treated with medication. We do not know whether the successful treatment of these conditions with medication will reduce the rate of violence, but it seems likely.

Medical Treatments and Ethical Problems

Only one medication, the adrenergic beta blocker Propranolol (Inderal), has been prescribed specifically to counteract violent behavior in traumatically brain-damaged individuals and others whose injury was followed by expressions of violence. It has also been used with some beneficial effect in children whose behavior is characterized by intermittent outbursts of violence following relatively minor frustrations.[6] Propranolol has been beneficial in treating violence in patients with dementia, personality disorders, PTSD, schizophrenia, mental retardation, autism, and brain injury. Use of this medication is considered symptomatic therapy, directed not at a psychiatric or neurologic illness per se but rather at violent behavior itself.

The mechanism underlying the purported effectiveness of this medication may relate to the lessening of the expression of sympathetic nervous system activity that normally activates an individual's fight-or-flight response to threat. The category of drug to which Propranolol belongs has been prescribed for a disparate range of conditions, from high blood pressure to angina, migraine prevention, and to lessen stage fright suffered by performers. The side effects of this drug include inducing or worsening depression, inducing sexual dysfunction in males, worsening bronchial asthma, and blocking the symptoms caused by low blood sugar. For this reason, it cannot be used by asthmatics or by diabetics who are taking insulin. Despite these caveats, the usual side effects are mild, and the range of effectiveness of this medication is extensive.

Though Propranolol has been used to control violent outbursts and shows great promise in this regard, the research data proving its effectiveness in reducing violent outbursts are rather thin. Hence, the drug has not yet been approved by the Food and Drug Administration (FDA) for this purpose. Practicing physicians can, of course, prescribe it for particular patients, but the limited FDA approval and the fear of lawsuits have limited its use in prisons and government institutions. Government employees, including physicians, seldom wish to expose themselves to criticism by taking unusual therapeutic initiatives. The research required to prove this drug's effectiveness in modifying violence would require considerable funding, which to date, has not been available.

Other studies of the effects of different drugs on violent behavior have been virtually ruled out, at least in the United States, by medical-ethical considerations. All subjects of medical research must provide informed consent of their own free will and cannot act under duress. Prisoners cannot exercise their own free will because they are in prison. Even if they volunteer for studies, the fact that they may be trying to curry favor as a way of possibly shortening their period of incarceration is regarded as a form of duress. These considerations have significantly restricted the introduction of promising therapeutic tools in the prison environment. These medical-ethical constraints seem unreasonable and unwarranted when the beneficiaries of successful treatment are the prisoners themselves. This utilitarian consideration and the potential health benefits to greater public are much more important than fairly abstruse theories about whether the prison environment constricts the exercise of will sufficiently to render a well-designed study unethical.

Another pharmacological study could focus on possible beneficial effects of medication prescribed to violent individuals diagnosed as having treatable psychiatric or neurologic conditions such as depression, mania, schizophrenia, epilepsy, or migraine. Any such study could generate accurate data only if it were performed in an institution in which the behavioral changes could be documented by guards, nurses, or other trained personnel. Otherwise researchers would be forced to depend on much less reliable, anecdotal evidence from the subjects, their families, and their friends. This kind of study is ethically defensible with regard to the recruitment of subjects diagnosed as having treatable neurologic and mental illnesses, but not with respect to controls. Because ethics require that all patients diagnosed as sick be treated, the rates of violence in an untreated control group cannot be compared to the rates of violence in a drug-treated group. Thus, as it is currently established, both the treatment of prisoners who may be sick and the nontreatment of prisoners who may be sick can be considered unethical. It is high time that practical reason cut this Gordian knot.

But many other practical and ethical problems must be considered. Individuals whose violence caused them to be incarcerated are easily identified, but they may not be violent in prison. Serial murderers, for example, tend to be model prisoners. The behavior of a man who killed his wife and three children during a period of stress may be exemplary throughout his life term in prison. If he were also suffering from a bipolar mood disorder, it would be very difficult to determine whether his lithium medication also had a beneficial effect on his violent outbursts if, in fact, these were not occurring anyway in prison.

How should treated subjects be handled in prison? Prisoners

who engage in violent behavior in prison frequently enough to measure the effect of treatment may pose problems of such severity to correctional officers that they are segregated in maximum security units and spend considerable periods of time in isolation. If their violent behavior is simultaneously controlled by isolation and other external restraints, it would be very difficult to determine the effect of their medication regime on violence directed at other inmates.

If the medication seemed to be effective, the subject might be transferred within the prison to a less intensely secure environment, but other inmates could be exposed to some undue risk of injury. In other words, the failure of therapy could be dangerous to other prisoners. Still, the potential for harm could be reduced, if not eliminated, by correctional officers, just as it is in most prisons now.

On the other hand, success might result in unforeseen danger. For example, I examined an extraordinarily violent, paranoid inmate with rapid mood swings and a history of unprovoked attacks on others. He was extremely and frequently violent, involved in several intense and dangerous fights each week. Often six trained guards were needed to subdue him, restrain him physically, and separate him from other prisoners.

This inmate was housed in the maximum security wing and was intermittently segregated in isolation cells. When housed in open units, the other inmates avoided him to the greatest extent possible, as he was widely regarded as dangerous and unpredictable. He was an unpopular figure and had mortal enemies among the inmates, many of whom bore the scars of violent encounters with him.

I suggested a trial of carbamazepine, an anticonvulsant that also effectively reduces mood swings. The medical staff at the prison found that this drug had a remarkably favorable effect on him. Several weeks passed without a violent outburst. The subject himself was delighted: for the first time in his life he felt that he could control his violent impulses. He no longer lost his temper suddenly or felt the need to viciously attack people who bothered him. But he confided to the prison psychiatrist that this benefit could affect his reputation within the prison, making him vulnerable to retaliation if the other prisoners realized that he no longer would explode against them with extremely violent attacks. Indeed, this prediction was borne out several weeks later, and he was beaten to unconsciousness, sustaining serious and permanent brain injuries in a fight initiated by his enemies who were no longer afraid to engage him in battle.

Despite this unfortunate outcome of my successful suggestion concerning therapy, it is usually beneficial to the prisoners if their psychiatric and neurologic disorders are diagnosed and treated. About a quarter of the inmates of U.S. prisons and jails have mental disorders.[7] The closing of large state mental institutions has shifted a large segment of the mental health population to the correctional system.

Several studies have shown lithium to be successful in reducing violent outbursts, first documented by Sheard et al. in their treatment of juveniles with the symptoms of bipolar affective disorder (manic-depressive illness) in a reform school.[8] Common clinical experience with anticonvulsants in treating mania leads me to believe that the older anticonvulsants like carbamazepine (Tegretol) and valproic acid (Depakote), as well as the newer ones like gabapentin (Neurontin) and lamotrigine (Lamictal), might be as successful as lithium in reducing violent behavior in previously violent individuals who also have bipolar affective disorder. Of the anticonvulsant medications, only valproic acid has been approved by the FDA for the purpose of treating bipolar affective symptoms. No study of anticonvulsants with respect to their efficacy in controlling violence has yet been conducted.

Depression and suicide are common among violent prisoners.[9] The use of antidepressants for treating depression is not scientifically controversial but has been attacked unfairly and in a sensational manner by those, including some physicians, who oppose the use of any medication that can improve mood, thought, or behavior. The very effectiveness of antidepressants seems to have sparked criticism of the use of fluoxetine (Prozac) and other selective serotonin reuptake inhibitors (SSRIs) like seratroline (Zoloft) and paroxitene (Paxil).

The SSRIs are remarkable drugs, effective in lifting depression in about 80 percent of patients, and, in higher doses, in lessening the symptoms of obsessive-compulsive disorder. They are remarkably free of serious side effects, although they tend to impair sexual response in both men and women and can affect appetite and cause drowsiness or insomnia. However, they are relatively expensive and have about the same spectrum of clinical activity as cheaper, older medicines such as desipramine and other tricyclic antidepressants (TCA). Tricyclics, on the other hand, have more side effects than SSRIs, including dry mouth, constipation, urinary retention, difficulty focusing the eyes, rapid heartbeat, sedation, memory loss, and confusion.

TCAs and SSRIs are about equal in effectiveness as antidepressants. Each has a three-week delay in the onset of the antidepressant effect after the initiation of therapy. TCAs have an advantage over SSRIs in that blood levels can be taken and doses can be adjusted to obtain a recognized therapeutic range. This serves as a guide for therapy and also as an important measure of the patient's compliance—an unexpectedly low blood level usually means that the patient has not been taking the medication.

When used in patients with manic-depressive disorder, all antidepressants can ameliorate depression but some can also precipitate a manic episode.[10] For this reason, mood stabilizers are preferable or should at least be used in combination with antidepressants. Antidepressants are the drugs of choice for patients who only have periods of depression without mania (unipolar depression) and obsessive-compulsive disorder.

Both mania and depression can precipitate violence, and intermittent violence can be correlated in certain individuals with either highs or lows of mood, or both, in bipolar patients. Mania often induces something akin to cocaine or amphetamine intoxication—excitement, irritability, lack of concern for the outcome of unwise actions and unwise statements—and also can cause intense paranoia. Depression can evoke anger, irritability, self-destructive actions, and a kind of "what's the use" attitude and lack of concern for outcomes. Though it is likely that the successful treatment of depression can reduce the level of violence in depressed patients, no formal assessments have been conducted.

In certain murderers, especially serial murderers, there is an obsessive-compulsive quality to their acts. Often their victims are alike in gender and age, the murders are committed in the same way, and the methodology they use to entrap and kill victims is also quite similar, time after time. Conceivably a medication that relieved obsessive thoughts could provide a benefit here, but it would be impossible to prove this logical conclusion scientifically.

Schizophrenia can also precipitate violence. In one study, the severity of schizophrenia, as evident from the degree of schizophrenic symptoms and low neuroleptic serum levels, was an important predictor of inpatient physical assaults on staff in a mental hospital.[11] A high prevalence of paranoid schizophrenia among men convicted of homicide (11 percent) and arson (30 percent) was reported by Taylor and Gunn, who consider even these high figures to be underestimates.[12] In a study of 203 prisoners remanded for psychiatric evaluation, a subgroup of 121 were psychotic. Of these, 23 percent were considered expressly driven by their psychotic symptoms to commit offenses. Those who were driven to offend by delusions were most likely to be seriously violent.

The treatment of schizophrenia with neuroleptics is effective in reducing delusions and hallucinations and would surely be indicated in violent schizophrenics, but most of the neuroleptic medications used to treat schizophrenia have many undesirable side effects. One of these is sedation. Turning a patient into a "zombie" is not a legitimate therapeutic goal. The long-term use of neuroleptics can result in the development of ugly involuntary movements of the mouth, face, and neck called tardive dyskinesia. This and other motor problems have limited the utilization of neuroleptics and make them ethically difficult tools for clinical therapeutic trials concerning the control of violence.

If epilepsy can cause violence, the standard treatment for it might reduce violence. This simple hypothesis is also controversial, because there is no clear consensus on the relationship between epilepsy and violence. But epileptic seizures and violence could theoretically be related in several ways:

1. Directed violence could occur during the automatism of a complex partial seizure.
2. Directed violence could be an outgrowth of an encephalopathy associated with the period of confusion that follows a seizure (the postictal state) or with the antiepileptic drug used to treat seizures.
3. Brain damage that predisposes an individual to violence might also cause seizures.

Although there have been many reports of an association between complex partial seizures and interictal violence, other studies that have failed to demonstrate any association between seizures and violence. The contradictions in these results may be related to varying definitions of epilepsy, violence, and aggression and to the selection of the patients who were studied. I seriously doubt that murder is ever the involuntary product of a seizure, although an individual could perhaps become violent during the confusional state that may follow a convulsion. The fact that brain damage can cause epilepsy and independently disinhibit behavior accounts for the higher prevalence of epilepsy in violent prisoners.

My view has moved close to that of Stevens and Hermann,[13] who contend that brain damage, not epilepsy, increases the chances of violent behavior. Brain damage, especially in limbic areas, can cause paranoia, and frontal damage can cause disinhibition. Paranoia and disinhibition are significant precipitators of violence, especially when combined with a history of childhood abuse. Limbic and/or frontal damage can also cause seizures, but seizures themselves rarely cause violence. Though the presence of seizures can be indicative of brain damage, it is the brain damage, not the seizures, that disinhibits.

It is highly unlikely that any pill would ever make a murderer "safe." Certainly, no such transformation has been demonstrated to date. Treatment might lower the risk of violence in the predisposed, but it will not eliminate it. I would strongly oppose the release of a murderer from prison based solely on an apparently successful medication regime. Nevertheless, drug treatments can be beneficial in the prison environment.

In my view, child abuse is the most important element in generating base instincts and the one that is most amenable to correction. The benefits to society of eliminating this cause of the impulse to act violently would be felt in lower rates of assault and murder and would break the devastating cycle of child abuse in succeeding generations. It might also decrease terrorist acts, dampen the will of citizens to fight aggressive wars, and reduce the number of hate crimes associated with racism.

Not less importantly, elimination of abuse would enable individuals to achieve their full potential. The brains of childen would develop and grow unimpeded by the burden of the base instincts that abuse engenders. No longer victimized, they would be less likely to be perpetrators of violent crime and could enter the portals of normal society rather than the gates of prison.

APPENDIX

Tools of Diagnosis: History, Physical Examination, and the Role of Tests

When I evaluate a person with any form of disordered behavior, I first look for evidence of brain impairment. The main tools that I use are the medical history and the physical neurological examination. Examining violent people in custody takes about three hours. In an ordinary clinical situation, the neurological evaluation of a new patient takes about an hour. The chief source of information is usually the patient himself. I make every effort to corroborate the history as the reports of the patient may be incomplete or inaccurate.

I begin by asking about the patient's birth. Sometimes parents have told their children about being a "blue baby" or having serious problems at or around the time of their delivery. I ask about accidents and head injuries, car crashes, falls on stairs and from trees, sports injuries, and the use of medicines or drugs that can damage the brain or affect behavior. I ask about headaches, epileptic seizures, dizzy spells, and fainting. My evaluation includes a review of any available medical, school, and police records. I also try to interview friends and family. It is critically important to determine if the patient was abused in childhood.

To determine whether abuse occurred, questions must be carefully framed about physical and sexual abuse. It is best not to pose a torrent of questions from a long checklist that can be answered "Yes" or "No." Lists of questions are invitations to falsely negative responses. The questions asked must be tailored to the answers received. Moreover, I have found it more fruitful to ask indirectly about how the parents responded in particular situations. For example, I ask what sorts of things the patient did as a child that led to a punishment. This approach puts the "blame" on the child rather than the parent. Violent individuals have a strong bias toward preserving their good opinion of their parents and often do not wish to blame them for abuse or to characterize them as abusers. For example, one person who had just said he had never been abused told me he was once severely punished for running away to avoid discipline for having broken a window when he was six. His sister and father immobilized him and burned the soles of his feet with a lighted candle to prevent him from running away again. In his opinion, this punishment fit the crime. He considered it to be reasonable and not abusive.

Another telling indicator of abuse is bed-wetting. Many abused children continue to wet their beds at night until adolescence. The physiological reason for this is not clear, but bed-wetting is one manifestation of stress. The response of parents to this behavior can open a line of very informative discussion about abuse. Some children are beaten daily for wetting the bed or are humiliated by such disciplines as being forced to wear stinking, urine-soaked clothes to school, being tied to a post at home like a dog, or recording bed-wetting on a calendar so the whole family and visitors can see whether the child wet his bed the night before.

Other childhood behaviors are also hallmarks of abuse, like fire-setting and cruelty to animals. When there was a fire in the house, who was thought to have set it? This is important because virtually every child who sets fire to his bed or that of his parents has been sexually abused. Victimizing helpless animals is also a way a child can direct his feelings of hatred and his desire to be in control without fear of retaliation. The physiology of this association is not clear but the association of these behaviors with abuse has been empirically proven.

"Who was the main disciplinarian?" and "What were you beaten with when you were spanked?" are very useful questions. Even violent inmates who have forgotten severe abuse may remember what they consider to have been good parental practices. Some of these practices are clearly abusive by my definition, like the use of a belt or wooden instrument directed elsewhere than their buttocks, spanking with the buckle of the belt, breaking of the skin, and punches to the face.

Being locked in a closet for an hour or more or in a room for a week can be quite terrifying to a child. Uncovering these extreme punishments raises other questions, like "What was the punishment for leaving the closet or the room before you were given permission? What happened if you tried to run away during a beating?"

I always ask about the worst punishment the person ever received for misbehaving. Some patients have told me very disturbing stories. Some children are forced to kneel on dry rice and salt for an hour; the more the child moves to try to relieve his pain, the more grains dig into his raw, bleeding skin, leaving permanent and verifiable scars. Abusive parents use electric wiring, broom handles, and other cruel and inappropriate tools to beat their children. Some parents even use lighted cigarettes to burn their children or hold their hands over an open flame as punishment.

The circumstances in which physical scars were sustained provide a window into the world of the violent individual. Linear scars and round scars on the back are usually caused by whips and cigarettes and cannot be self-inflicted. Almost every normal person remembers the incidents that have caused scarring on the portions of his body visible to him and can proudly tell the stories of how the scars were sustained. But frequently, violent people cannot identify the causes of many of their scars,which are clearly the result of burns, knife wounds, bullet wounds, and other trauma.

Such memory lapses suggest that the person may be engaging the kind of psychological mechanism for forgetfulness that is used in dissociation and that bespeaks severe abuse. This is particularly telling when the cause of scarring is stated in the medical records.

When I ask about sexual abuse, I try to make it sound as if many people are sexually abused and that it is not a big deal. I often start by saying we know from watching television programs like *Oprah* that a lot of children are asked to do sexual things for grown-ups. Then I ask, "Who did that to you?"

Because the anal penetration of children damages the rectum and colon, common symptoms in victims include lower abdominal pains of unknown cause, painful bowel movements, bloody bowel movements, and constipation. Asking if the person has ever had these symptoms is accepted as the ordinary questioning of a family doctor. A similar approach is taken to determine whether sexual abuse of boys was inflicted by putting objects into their penises. As the memory of such experiences may be repressed, it is often more revealing to ask, "Did you ever have blood come out of your penis? Did you ever have bladder infections?" These questions are interspersed with benign questions about nosebleeds, earaches, rashes, and so on.

The family may provide convincing details of abuse that the patient has forgotten. For example, one convicted rapist-murderer on death row, who did not remember his crime, denied ever being sexually abused as a child. However, his older sister testified that she had seen him being repeatedly anally penetrated when he was eight by their uncle, in whose home their family was living. She described her brother crying and screaming while he tried to escape from his uncle. Although the rapist-murderer recalled that he did not like his uncle, he did not remember being raped by him.

Terrible stories of abuse told by defense witnesses at sentencing trials can move juries and persuade judges not to impose a death sentence, so the defendant has an important stake in showing that mitigating factors exist. Conventional wisdom among prosecutors and society at large holds that the fabrication of stories of abuse occurs often. But in twenty-five years of seeing the most violent people in America, I have only once encountered an inmate whose claim of abuse was reliably refuted by other family members. In my experience, falsification has weighed heavily in the other direction. I have seen many murderers whose claim not to have been abused was contradicted by independent evidence.

Many condemned prisoners would literally go to their deaths rather than consciously and publicly describe their abuse by their parents. Many families of the condemned would much rather see their relative executed than reveal the story of abuse that implicates them as either perpetrator or fellow victim.

I perform a physical neurological examination to determine whether various parts of the brain are functionally intact. The human brain is organized in layers of interacting regions. It is easiest to test the lower, more primitive parts of the brain. At the base, located just behind the mouth, is the brain stem, which controls the most elemental functions—breathing, blood pressure, and swallowing. Above that, behind the nose and its sinuses, is the diencephalon. It controls body temperature, appetite, sleep, and wakefulness. Circling around that, in the central part of the brain, is the limbic system and amygdala-hippocampus complex, the seat of the primitive emotions involved in sexual behavior, fear, anger, attack, and memory. Covering the entire surface of the brain and lying just below the skull is the cerebral cortex, the thick layer of gray matter (nerve cells) that is the source of conscious mental processes.

The greatest development of the cortex in the animal kingdom is seen in humans. The cortex is anatomically separated into regions called lobes. Vision is controlled by the occipital lobes at the back of the head; somatosensory information like touch, pain, and position in space is interpreted behind and above the ears in the parietal lobes; smell and taste are determined by the temporal lobes at the sides of the brain near the ears. Speech and the understanding of language comes from the left posterior temporal lobe; voluntary movement from the motor strip, the posterior part of the frontal lobes that lies just in front of the ears. The left side of the brain controls speech and the right side of the body. The right side of the brain controls the left side of the body. Voluntary movement is modified by deep subcortical centers like the basal ganglia and the cerebellum. Almost all vertebrates have well-developed occipital, parietal, and temporal lobes and motor control systems that are similar to humans.

Thus in my neurological examination, I test reflexes and sensorimotor functions, comparing the patient's right and left sides, and observe his stance and gait. I use a tape to measure the head circumference.

I also check gross motor coordination. I ask people to skip, to hop, and to walk a straight line forward and backward. I ask them to spread apart their fingers and hold their hands apart and I look for choreiform movements—discontinuous, involuntary, little jerks of the fingers and arms. I check for mixed dominance: if a person is right-handed but left-footed, for example, it might mean that his nervous system has not developed the way it should. Abnormality on a single test, by itself, may be insignificant, but a pattern of abnormalities does indicate that the brain is malfunctioning. Asymmetrical (lateralized) abnormalities are most likely to reflect structural abnormalities of the central nervous system, but severe damage to the brain can be present in patients whose sensorimotor functions, coordination, and reflexes are normal.

The brain is the organ of cognition, so cognitive functions must be tested too. I begin with a brief, formal psychological battery called the Mini Mental Status Examination, asking the patient to state the day, date, time of day, and whereabouts; to name several objects; and to repeat three words immediately. I determine if he can subtract seven from one hundred, and can perform four more serial subtractions of seven. I ask him to spell the word "world" forward and backward. I have him read, repeat, and write; copy a standard diagram; and follow a three-step command; and I ask him to recall the three words that I instructed him to remember earlier in the examination. These tests of orientation, memory, calculation, reading, writing, and speaking are very useful for uncovering evidence of damage to the two-thirds of the brain that are located in the back (the parietal, temporal, and occipital lobes), but they tell very little about the frontal lobes. Located behind the forehead, above the eyes, and extending back to the ears, the frontal lobes are the most difficult region of the brain to assess clinically.

The frontal lobes of humans are massive in relation to the other parts of the cortex and account for about one-third of the cortical mass.[1] This proportion is highest in humans and distinguishes the human brain from all other primates. It is the job of the frontal lobes to focus attention and to modify and inhibit the behavioral impulses that surge up from the other parts of the brain. The frontal lobes permit us to say to ourselves, "No. Don't say (or do) that," in response to our urges. The frontal lobes subserve judgment and allow us to anticipate adverse or favorable circumstances, to plan, invent, modify, and adapt in response to changes in our environment. Frontally damaged individuals cannot focus their attention for long periods of time. They may be withdrawn, apathetic, irritable, and unable to control the expression of emotions such as anger and sexual attraction. People with frontal damage may express desires that their intact frontal lobes would have controlled. They might spend too much money, drive at excessively high speeds, or fight. Frontally damaged people often cannot keep their behavior within the general rules of society.

With respect to behavior and variability, being frontally damaged is like being drunk. Some people become boastful, others irritable, still others calm and apathetic. Some have no interest in sex, others are hypersexual. They may be argumentative, generous, foolish, happy, or sad. Considerable fluctuations in mood and behavior can occur in a short time, unmodulated by the social brakes that the frontal lobes normally provide. When people suffer from psychiatric diseases that ordinarily cause mood swings, the mood swings are wider and occur more rapidly in the presence of frontal damage.

Frontal lobotomy was once the most widely used somatic therapy used in psychiatry. Egaz Moniz, the neurosurgeon who introduced it for treating psychiatric disorders, won a Nobel Prize. Ultimately, it was found to worsen schizophrenia and mania, although it could improve some obsessions and phobias by inducing indifference. It has almost no role now in the treatment of psychiatric disorders because its effects are unpredictable and permanent.

Despite their importance, the frontal lobes can be damaged or even surgically removed without causing any abnormality in other particularly human cognitive functions such as speech, arithmetical ability, reading, writing, and memory. Tested by the Mini Mental Status Examination, these functions often remain intact after frontal lobe injury, but profound and devastating changes can occur in the social life of frontally damaged people.

It is thus possible for a person to become a social imbecile because of frontal damage and still have a normal Mini Mental Status Examination and a normal IQ with the ability to speak, read, write, calculate, and remember. These functions are all primarily carried out by the back part of the brain, the part behind the ears (the parietal, temporal, and occipital lobes). Therefore, one of the most important parts of the neurological examination in people whose behavior has been violent is the assessment of the frontal lobes, which direct judgment and the executive ability to deal with complexity. Unfortunately, there is no reliable objective standard or test for measuring judgment. However, there are several ways to search for evidence of deficits in other frontal lobe functions.

When I tap repeatedly on the bridge of someone's nose, it is normal for him to blink two or three times in response to a potential threat to his eyes. When it becomes clear that there is no threat, the blinking should stop. If it does not, it indicates that he cannot accommodate: he cannot suppress the urge to blink even though there is no threat. He cannot adapt to a new situation, an inflexibility that reflects frontal dysfunction.[2]

The word-fluency test also examines frontal function by testing the ability to generate words out of a context. In this test, I ask the person to name as many words as he can in sixty seconds that begin with the letter "F." A normal score is fourteen, plus or minus five; it is abnormal to name less than nine words. This is not an intelligence test—a person with frontal damage might do as well as anyone else on an IQ test or other measurements involving reading, writing, calculating, and memory. But because frontally damaged people cannot improvise or succeed in using old knowledge in a new way, the word-fluency test provides an index of frontal dysfunction.[3]

Paying attention to two sensory stimuli simultaneously can be difficult for frontally damaged people. They tend to disregard the stimulus that is futher from the face. For example, if I touch his foot and hand, the patient with his eyes closed may report being touched only on the hand. If his face and hand are simultaneously touched, he may report having been touched only on the face.[4]

Testing the nucocephalic reflex determines the ability to adapt to a postural change, so it is often helpful in diagnosing frontal disease. With the patient's eyes closed, I place my hands on his shoulders and quickly turn him to the right or to the left. Normally, the head follows the shoulders, the frontally impaired may not move their heads: they cannot adapt to this postural change.

In the antisaccade test, I face a patient, holding my hands to my sides, I alternately move my right and then left index fingers. After directing the patient to look toward the side on which my finger moves, I then ask him to look toward the side that does *not* move. Because patients with frontal damage have trouble suppressing the urge to look at the moving finger, this test uncovers impulsivity and inattentiveness.[5]

Frontally damaged individuals often cannot perform two- and three-stepped motor sequences repetitively even though their strength and coordination are normal. Therefore, complex repetitive motor tasks are sensitive indicators of frontal function. Using the three-stepped Luria test, I instruct the patient repeatedly using the precise sequence of striking his thigh with his right palm, fist, and the side of his hand. Then this is repeated on the left. Next, the person is asked to strike his thighs simultaneously with the right palm and the left fist, repeatedly, but alternating with each blow the fist and the palm. This is a two-stepped Luria test. Abnormality on these tests indicates frontal dysfunction.

The frontal lobes also control voluntary eye movement. To test this, I ask a person to look at and to follow my finger with his eyes as I slowly but steadily move it in a forty-five-degree arc to the right and to the left. If his eyes can only follow it in brief discontinuous jerks, or he cannot maintain his gaze on my finger, his frontal eye fields are not working properly.

Upward gaze is another frontal function. When asked to look upward, the individual should be able to move his eyes upward by five millimeters. Also, a person should be able to stare at one object to his side without moving his eyes or to stick out his tongue with his eyes closed and maintain this for thirty seconds. If he cannot, it represents motor impersistence, another sign of frontal dysfunction.[6]

When I instruct them to relax, patients with frontal disturbances often allow their limbs to remain suspended in the air when I raise one leg or an arm and let go. This abnormality is called paratonia. When this sign is severely abnormal, the limb will remain in any position in which it was placed. This is called waxy flexibility or catatonia; it can also be encountered in severe psychiatric disturbances like schizophrenia.[7]

Abnormal results on one of these tests is not necessarily indicative of frontal damage, but a pattern of abnormality—two or three abnormal tests—can sustain this diagnosis. Three abnormal tests reliably predict abnormalities on both the MRI of the brain and on neuropsychological testing using the Halstead-Reitan battery. The most sensitive and specific of these frontal lobe neurological tests are the inability to perform smooth visual pursuit and the inability to perform the three-stepped Luria test.[8]

The main life problems that frontal damage creates for people are the loss of social inhibition and social judgment. A normal person who ignores tapping on the bridge of his nose by a physician can brush off being jostled in a crowd or having his foot stepped on or being honked at by someone in a car. He might ignore or respond to an angry schoolmate who says "I'll get you" by seeking the assistance of a teacher, principal, or bus driver. Normal people can screen out and evaluate the relative importance of aspects of their environment, dismissing the irrelevant. Frontally damaged people, however, live on the edge: they can lose control at any time.

An interesting study evaluated veterans who had suffered penetrating traumatic brain injuries during the Vietnam War.[9] Veterans with frontal damage were compared to those with damage elsewhere in the brain. The study revealed that the veterans with frontal damage were several times more likely to behave violently in the years following their recovery from the acute injury, though only a minority of those who had sustained frontal damage were violent.

Explosive, uncontrollable rage often is expressed by a person who no longer has the neurological capacity to moderate the primal, limbic feelings of fright and the urge to fight. Poor judgment in carrying out a plan, like expressing sexual feelings for a child, carrying a knife to school, or raping and killing in an unsuccessful robbery can also reflect frontal dysfunction.

In the majority of murderers that I have examined, there is objective evidence of neurological abnormalities, primarily in the frontal lobe. Two-thirds of the thirty-one murderers I examined sequentially over a five-year period had abnormal frontal signs on the physical examination, virtually all had abnormal psychological testing, almost half had abnormal electroencephalograms (EEG), and almost half had abnormal MRIs.[10] Murderers have commonly been exposed to factors that are known to damage the brain, such as prenatal fetal alcohol syndrome and childhood traumatic brain injury.

The EEG, which measures the sum of electrical activity in the surface of the brain the way an electrocardiogram (EKG) does for the heart, has been used a great deal in both psychiatry and neurology. In the presence of paroxysmal episodes of spikes, or spikes and waves, the EEG is an excellent discriminator of patients having epileptic seizures as opposed to patients exhibiting hysterical pseudoseizures. However, because the EEG is likely to be performed at a time when the patient is not having seizures, it cannot rule out epilepsy.

Over the past few years, EEG use has increased. Sometimes it involves prolonged video monitoring of the patient, who is generally confined to one room over periods of time varying from twelve hours to two weeks. More common is ambulatory telemetry, where electrodes are connected to a tape recorder carried by the patient. An event marker can be activated by the patient or a caretaker. A computer "reads" the tape and prints out abnormal sections of the EEG as well as those marked by the event recorder. The lengthy records of EEG telemetry reduce the sampling problem posed by the routinely scheduled hour-long EEG. Prolonged recording also includes several hours of sleep, which is particularly valuable in uncovering an epileptic tendency.

Normal EEG results do not exclude the existence of a brain disorder. Scalp electrodes may make it difficult to localize symptoms because those originating in deep structures, such as the mesial temporal areas, may not be picked up by the surface electrodes. Even depth electrode recording may be falsely negative when the placements are not in exactly the areas associated with firing. Sleep EEG records significantly increase the likelihood of finding EEG abnormalities, particularly in focal areas, such as temporal lobe lesions. Special electrode placements, such as sphenoidal and nasopharyngeal electrodes, have been useful to a limited degree.

Advances in EEG technology may ultimately effect changes concerning its use. This is particularly so in the context of computerized EEG monitoring, which allows quantitative analysis of the distribution of wave forms in various parts of the brain. This technique was overused, and early claims that it could diagnose mental illnesses were proven incorrect. The American Academy of Neurology has issued a position paper discrediting the diagnostic use of the technique, but it can provide a very useful and objective reading of the EEG.[11] Other uses of computerized analysis of the EEG involve evoked potentials, of which the correlation with of clinical conditions is still in the experimental stage.

Computerized tomography (CT) is a computer-assisted X-ray examination that produces images of the brain. It is useful for detecting calcification, hemorrhages, and large strokes. Magnetic resonance imaging (MRI) is generally superior to CT in demonstrating abnormalities of the brain. A computer generates the image by analyzing energy waves that are initiated by a powerful electromagnetic field. Like a child's magnet with iron filings, the magnetic field around the brain polarizes molecules, chiefly hydrogen. The most plentiful source of hydrogen is water (H_2O). The hydrogen atoms in H_2O carry a positive charge and are oriented toward the negative side of the electromagnetic field. When the orientation of the electromagnetic field is shifted, the orientation of hydrogen atoms also shifts, creating a wave of energy; its strength is proportional to the amount of water in the part of the brain that is being imaged. Computerized analysis of this energy wave can create a series of images of the brain, like photographs of a tomato that has been thinly sliced.

MRI is especially useful for diagnosing strokes, tumors, multiple sclerosis, and vascular malformations. It is vastly superior to CT for imaging the spinal cord. With a variety of software programs, the MRI can provide arteriograms and venograms, thereby outlining the arteries and veins of the brain without the need for the dangerous injection of dyes. MRI sensitively detects swelling (edema) and the blood products of previous hemorrhages (hemosiderin).

Despite these advantages over the CT, both CT and MRI show normal results in many clinical conditions in which the brain is very abnormal. The combination of abnormal brain and normal MRI and CT exists in dementing illnesses like Alzheimer's, intoxication, epilepsy, Parkinson's (and other movement disorders), mental retardation, autism, learning disorders, dyslexia, schizophrenia, mania, depression, anxiety disorder, and personality disorders.

The MRI is so wonderful that it is easy to forget that it has not been standardized. Interpretation depends solely on the experience of the radiologist who is reading it. No published norms exist for MRI that address crucial questions: What is the normal volume of the frontal gray matter? How much smaller than average must this number be to be abnormal? Is the proportionate size of gray and white matter in the temporal lobe normal? Again, what degree of deviation from the average is normal? Normative data is not generally available, though Adrian Raine, using quantitative data, found that frontal gray matter was significantly smaller in men with antisocial personality disorder than in controls.[12]

New techniques use isotopes and MRI to obtain images of brain function. Positron emission therapy (PET) and single photon emission computerized tomography (SPECT) are techniques that depend on isotopes to reflect cerebral blood flow. The amount of blood flow to various regions of the brain depends upon the level of cerebral activity. In general, gray matter is more metabolically active than white matter and receives more blood; portions of the gray matter that are physiologically active receive more blood flow than gray matter that is inactive. In other words, the cognitive portions of the brain receive more blood flow when the subject is reasoning.

Functional magnetic resonance imaging (fMRI) provides similar information to that of PET and SPECT about cerebral blood flow and, by implication, brain activity. fMRI does not utilize injected isotopes or radiation and is therefore, safer, less expensive, and ultimately more available. PET, SPECT, and fMRI techniques are currently are currently the subjects of investigation, but their use as tools for the investigation of the brain should become more common in the future.

Fundamentally, the functional tests of the brain are misused when the radiologist performing them asks the subject to lie quietly, thinking of nothing. The radiologist cannot know if the subject is obsessing over a remark he had made to his mother-in-law the day before, ruminating about escaping from custody, feeling depressed, anxious, manic, or performing calculations to project next year's income. Each thought and mood utilizes a different part of the brain and increases or decreases regional brain activity and blood flow.

Often defense lawyers order a scan to show an abnormal brain. Prosecutors want to see a normal brain. The PET, SPECT, and fMRI scans of brains supposedly at rest cannot provide that information. The brain must be working in a standard way for the results to be useful. For instance, a continuous performance test typically engages the frontal lobes, whereas reading engages the left parietal lobe. Using PET, SPECT, and fMRI to investigate frontally damaged criminals is futile unless they are requested to activate their frontal lobes with a psychological test. Damaged subjects will fail to activate. When dyslexics are asked to read, it is not the left parietal lobe that "lights up" but other parts of the brain.[13] But functional scanning of dyslexics who are unengaged in a reading task is likely to be normal.

Any scanning laboratory that uses these tests must develop its own normal control values. Techniques vary so much that one laboratory cannot rely on another laboratory's normal standards.

The mind and the brain can be disordered by conditions whose presence can only be detected by examinations of body fluids, namely, endocrine, infectious, metabolic, and toxic disorders. These disorders require investigation generally by blood tests such as complete blood count, sedimentation rate, urea, electrolytes, glucose, toxicology screen, syphilis, HIV, liver, and thyroid functions. Measurement of lactate B_{12}, folate, calcium, phosphorus, magnesium, and cortisols, porphyrin screens, heavy metals, antinuclear antibodies, carbon monoxide, and other tests may be required as clinically indicated. Lumbar punctures have become less necessary as imaging has improved, but inflammatory conditions of the brain still require cerebrospinal fluid examination for diagnosis.

The patient with possible brain damage should be evaluated in detail using the conventional medical approach. There is no substitute for an adequate history. Evaluation should include not only a full physical but also conventional neurological and psychiatric examination. The history and examination usually provide the diagnosis. This can be consolidated and extended by means of neuropyschological assessment, appropriate blood and urine tests, waking and sleeping EEGs, and neuroradiological investigations.

Notes

Prologue

1. "Defense lawyer David Bruck argued during the trial that the two boys died during a botched suicide attempt by a woman whose troubled life caused her to succumb to major depression. Defense witnesses testified about Ms. Smith's father's suicide when she was six, her step-father's molestation and her two suicide attempts." Molly McDonough, "Susan Smith Gets Life in Prison," *Spartanburg (S.C.) Herald-Journal*, July 29, 1995.

1. The Theory of Violence as Taught by Louis Culpepper

1. Except for well-publicized cases in the public domain, the names and locations of all the subjects in this book have been changed.

2. Murder on the School Bus

1. This case was reported by Dr. Dorothy Lewis, using a different name, in her book, *Guilty by Reason of Insanity* (New York: Ballantine, 1998).
2. M. Wolfgang, "Delinquency and Violence from the Viewpoint of Criminality," in *The Neural Bases of Violence and Aggression*, W. S. Fields and W. H. Sweet, eds. (St. Louis, Mo.: Warren Green, 1975), pp. 456–93.
3. S. S. Shanok and D. O. Lewis, "Medical History of Abused Delinquents," *Child Psychiatry and Human Development* 11 (1981): 222–31.
4. Unlike generalized grand mal seizures—during which people lose consciousness, collapse, and shake—complex partial seizures produce much more subtle symptoms: clouded consciousness, confusion, inappropriate behavior. Partial seizure episodes are characteristically brief and nonviolent, and the person usually does not remember what happened during the seizure.
5. There is a peculiar link between migraine and violence. Thirty-nine percent of the thirty-two fifteen-year-old delinquents I examined had migraines and only 6 percent of thirty-two nondelinquents, matched for age, sex, race, and socioeconomic level, a highly statistically significant difference, but one that is as yet unexplained. D. O. Lewis, J. H. Pincus, R. Lovely, et al., "Biopsy-

chosocial Characteristics of Matched Samples of Delinquents and Non-Delinquents," *Journal of American Academy of Child and Adolescent Psychiatry* 26 (1987): 744–52.

6. J. Grafman, K. Schwab, D. Warden, et al., "Frontal Lobe Injuries, Violence, and Aggression: A Report of the Vietnam Head Injury Study," *Neurology* 46 (1996): 1231–38.
7. C. George, et al., "Social Interactions of Young Abused Children: Approach, Avoidance, and Aggression," *Child Development* 50 (1979):306–18; M. H. Teicher et al., "Preliminary Evidence for Abnormal Cortical Development in Physically and Sexually Abused Children Using EEG Coherence and MRI," *Annals of the New York Academy of Sciences* 82 (1997):160–74; J. D. Bremner et al., "Magnetic Resonance Imaging–based Measurement of Hippocampal Volume in Post-traumatic Stress Disorder Related to Childhood Physical and Sexual Abuse," *Biological Psychiatry* 41 (1997):23–32.
8. C. S. Widom, "The Cycle of Violence," *Science* 244 (1989): 160–66.
9. D. Freedman and D. Hemenway, "Precursors of Lethal Violence: A Death Row Sample," *Social Science Medicine* 50 (2000): 1757–70. B. Rivera and C. S. Widom, "Childhood Victimization and Violent Offending," *Violence Victim* 5 (1990): 19–35. M. G. Maxfield and C. S. Widom, "The Cycle of Violence Revisited Six Years Later," *Archives of Pediatric and Adolescent Medicine* 150 (1996): 390–95.
10. Recently this has been shown to be abnormal, a manifestation of paranoid thinking, even in disadvantaged environments. S. L. Bailey, R. L. Flewelling, and D. P. Rosenbaum, "Characteristics of Students Who Bring Weapons to School," *Journal of Adolescent Health* 20: (1997): 261–70.
11. J. Modestin, "Criminal and Violent Behavior in Schizophrenic Patients: An Overview," *Psychiatry and Clinical Neuroscience* 52 (1998): 547–54. M. I. Krakowski, A. Convit, J. Jaeger, et al., "Inpatient Violence: Trait and State," *Journal of Psychiatric Research* 23 (1998): 57–64. P. Lindquist and P. Allebeck, "Schizophrenia and Crime: A Longitudinal Follow-up of 644 Schizophrenics in Stockholm," *British Journal of Psychiatry* 157 (1990): 345–50.
12. According to the Socialist Equity Party in England Robert Thompson lived with his alcoholic mother. His father, who had left home five years earlier, was also a heavy drinker who beat his wife and children. His seven boys would hit and abuse one another. One had voluntarily requested to be placed outside his parental home. Jon Venables's parents were also separated. His mother had psychiatric problems. Jon's school behavior was disturbed: head-butting walls, slashing himself with scissors, and hanging upside down on coat pegs. Both boys were frequently truant from school.
13. S. B. Guze, D. W. Goodwin, and J. B. Crane, "Criminality and Psychiatric Disorders," *Archives of General Psychiatry* 20 (1969): 583–91.

3. *Murder by Abuse*

1. D. O. Lewis, S. Shanok, J. Pincus, and G. Glaser, "Violent Juvenile Delinquents: Psychiatric, Neurological, Psychological, and Abuse Factors," *Journal of the American Academy of Child and Adolescent Psychiatry* 18 (1979): 307–19.
2. The clear and waterlike cerebrospinal fluid bathes the brain and spinal cord. Produced in natural cavities in the brain, it passes through special holes and circulates over the coverings of the brain before it is absorbed. If there is an obstruction to the flow of cerebrospinal fluid or if more fluid is produced than can be absorbed, pressure builds up in the brain, impeding normal functioning. This condition, known as hydrocephalus, can cause the brain to malfunction in cognitive capacity and motor performance. When hydrocephalus develops in the first year of life, before the skull has permanently closed, the increased pressure causes the head circumference to enlarge.
3. The cerebellum is a portion of the brain located near the base that has no cognitive function but modifies movement and helps coordination.
4. Standard psychological tests that primarily assess posterior (not frontal) cerebral functions. One section of the Halstead-Reitan Battery, Categories, assesses frontal (executive) functions.
5. The Rorschach is a series of inkblot designs that the patient is asked to interpret that reveal intellectual and emotional factors. The interpretations can be scored and classified in a standardized manner. The Rorschach is one method used to detect psychotic thinking.
6. C. S. Widom, "The Cycle of Violence," *Science* 244 (1989): 160–66.
7. B. Rosslund and C. A. Larson, "Crimes of Violence and Alcohol Abuse in Sweden," *International Journal of Addiction* 14 (1979): 1103–15; and T. Loberg, "Belligerence in Alcohol Dependence," *Scandinavian Journal of Psychology* 24 (1983): 285–92. P. Lindquist, "Criminal Homicide in Northern Sweden, 1970–1981: Alcohol Intoxication, Alcohol Abuse, and Mental Disease," *International Journal of Law and Psychiatry* 8 (1986): 19–37.
8. G. M. Asnis, M. L. Kaplan, G. Hundorfean, et al., "Violence and Homicidal Behavior in Psychiatric Disorders," *Psychiatric Clinics of North America* 20 (1997): 405–25.

4. *Genes, Geography, and the Generation of Violence*

1. P. A. Brennan, S. A. Mednick, and B. Jacobsen, "Assessing the Role of Genetics in Crime Using Adoption Cohorts," *Ciba Foundation Symposium* 194 (1996):115–23. S. A. Mednick, W. F. Gabrielli, Jr., and B. Hutchings, "Genetic Influences in Criminal Convictions: Evidence from an Adoption Cohort," *Science* 224 (1984): 891–94.
2. M. J. Gotz, E. C. Johnstone, and S. G. Ratcliffe, "Criminality and Antiso-

cial Behavior in Unselected Men with Sex Chromosome Abnormalities," *Psychological Medicine* 29 (1999): 953–62.

3. There is also a theory that elevated testosterone levels cause violence that does not hold up to close scrutiny; alcoholism and nonviolent "psychopathy" seem to account for most of the association. Lowering testosterone does not prevent sexual crimes. E. G. Stalenheim et al., "Testosterone as a Biological Marker in Psychopaths," *Psychiatry Research* 77 (1998): 79–88. G. C. Hall, "Sexual Offender Recidivism Revisited: A Metaanalysis of Recent Treatment Studies," *Consultations in Clinical Psychology* 63 (1995): 802–9. N. Heim and C. J. Hursch, "Castration for Sex Offenders: Treatment or Punishment? Review and Critique of European Literature," *Archives of Sexual Behavior* 8 (1979): 281–304. J. Raboch, H. Cerna, and P. Zemek, "Sexual Aggressivity and Androgens," *British Journal of Psychiatry* 151 (1987): 398–400. M. Richer and M. L. Crismon, "Pharmacotherapy of Sexual Offenders," *Annals of Pharmacotherapy* 27 (1993): 316–20. J. Zverina, J. Zimova, and D. Bartova, "Catamnesis of a Group of 84 Castrated Sexual Offenders," *Cesk Psychiatry* 87 (1991): 28–34. J. Zeverina, R. Hampl, and L. Sulocava, "Hormonal Status and Sexual Behavior of 16 Men After Surgical Castration," *Archivo d' Italiano Urologia Nefrologia et Andrologia* 62 (1990): 55–58.
4. J. S. Alper, ed., "Biological Influences on Criminal Behavior: How Good Is the Evidence?" *BMJ* 310 (1995): 272–73. All material in this paragraph and the next are from Alper's article.
5. As monoamine oxidase-A is involved in serotonin metabolism, some scientists suggested that the genetic defect might be related to older reports of low serotonin as a marker for violence. Paradoxically, however, the loss of monoamine oxidase-A leads to *increased* serotonin levels in the brain and other body tissues. H. G. Brunner, X. O. Breakefield, H. H. Ropers, and B. A. van Oost, "Abnormal Behavior Associated with a Point Mutation in the Structural Gene for Monoamine oxidase-A," *Science* 262 (1995): 578–80.
6. H. G. Brunner, oral communication, September 19, 1996.
7. In France, the rate was 1.1, in Britain, 0.5. One of the highest rates in Europe was in Finland with a rate of 3.2. Fox Butterfield, News of the Week in Review, *New York Times*, July 26, 1998.
8. Ibid.
9. R. E. Nisbett, "Violence and U. S. Regional Culture," *American Psychology* 48 (1993): 441–49.
10. U. S. Department of Justice, Bureau of Justice Statistics, 1998.
11. U. S. Department of Health and Human Services, Child Maltreatment 1996: Report to the States to the National Child Abuse and Neglect Data System.
12. U. S. Department of Health and Human Services, Child Maltreatment 1991: Report to the States to the National Child Abuse and Neglect Data System.

13. Quotes attributed to Evelyn Wise are derived from my notes and memory.
14. P. Y. Blake, J. H. Pincus, and C. Buckner, "Neurologic Abnormalities in Murderers," *Neurology* 45 (1995): 1641–47.
15. D. O. Lewis, J. H. Pincus, B. Bard, et al., "Neuropsychiatric, Psychoeducational, and Family Characteristics of Fourteen Juveniles Condemned to Death in the United States," *American Journal of Psychiatry* 145 (1988): 584–89. M. Feldman, K. Mallouh, and D. O. Lewis, "Filicidal Abuse in the Histories of Fifteen Condemned Murderers," *Bulletin of the American Academy of Psychiatry and the Law* 14 (1986): 345–52.
16. A. J. Sedlak and D. D. Broadhurst, U. S. Department of Health and Human Services National Incidence Study, September 3, 1996.
17. A. Mahoney, W. O. Donnelly, T. Lewis, et al., "Mother and Father Self-Reports of Corporal Punishment and Severe Physical Aggression Toward Clinic-Referred Youth," *Journal of Clinical Child Psychology* 29 (2000): 266–81.
18. T. V. Gurvits, M. W. Gilbertson, N. B. Lasko, et al., "Neurological Status of Combat Veterans and Adult Survivors of Sexual Abuse-PTSD," *Annals of the New York Academy of Science* 821 (1997): 468–71. Post-traumatic Stress Syndrome (PTSD) is a synonym for combat fatigue, combat neurosis, and shell shock. It refers to a mental state in which the subject's awareness is disassociated from his surroundings. It occurs in people subjected to situations in which their lives or corporal integrity have been threatened.
19. M. A. Kerr, M. M. Black, and E. Frercoeur, "Failure to Thrive, Maltreatment, and the Behavior and Development of Six-Year-Old Children from Low Income, Urban Families: A Cumulative Risk Model," *Child Abuse and Neglect* 24 (2000): 587–98.
20. National Center on Child Abuse and Neglect, 1993.
21. J. E. Richters and P. Martinez, "The NIMH Community Violence Project, I: Children as Victims of and Witnesses to Violence," *Psychiatry* 56 (1993): 7–21. P. G. Ney, T. Fung, and A. R. Wickett, "Causes of Child Abuse and Neglect," *Canadian Journal of Psychiatry* 37 (1989): 401–5. J. M. Leventhal, S. A. Egerter, and J. M. Murphy, "Reassessment of the Relationship on Perinatal Risk Factors and Child Abuse," *American Journal of Diseases of Childhood* 138 (1984): 1034–39. L. J. McIntosh, N. E. Roumayah, and S. F. Bottoms, "Perinatal Outcome of Broken Marriage in the Inner City," *Obstetrics and Gynecology* 85 (1995): 233–36. C. Holzman, N. Paneth, R. Little, et al. "Perinatal Brain Injury in Premature Infants Born to Mothers Using Alcohol in Pregnancy," *Pediatrics* 95 (1995): 66–73. S. R. Snodgrass, "Cocaine Babies: A Result of Multiple Teratogenic Influences," *Journal of Child Neurology* 9 (1994): 227–33. J. Bays, "Substance Abuse and Child Abuse: Impact of Addiction on the Child [Review]," *Pediatric Clinics of North America* 37 (1990): 881–904. R. Famularo, T. Fenton, and R. Kinscherff, "Medical and Developmental Histories of Maltreated Children," *Clinical Pediatrics* 31 (1992): 536–41. C. H. Leonard, R. I. Clyman, R. E.

Piecuch, et al., "Effect of Medical and Social Risk Factors on Outcome of Prematurity and Very Low Birth Weight," *Journal of Pediatrics* 116 (1990): 620–26. M. Klein and L. Stern, "Low Birth Weight and the Battered Child Syndrome," *American Journal of Diseases of Childhood* 122 (1971): 15–18. J. E. Oliver, "Intergenerational Transmission of Child Abuse: Rates, Research, and Clinical Implications [Review]," *American Journal of Psychiatry* 150 (1993): 1351–24. M. P. Janicki and J. W. Jacobson, "The Character of Developmental Disabilities in New York State: Preliminary Observations," *International Journal of Rehabilitation Research* 5 (1982): 191–202. G. D. Wolfner and R. J. Gelles, "A Profile of Violence Against Children: A National Study," *Child Abuse and Neglect* 17 (1993): 197–212.

22. Any theory that proposes to explain the rise in violence in the thirty years between the 1960s and the 1990s must also address the drop in homicide rates during the 1990s. Many attribute the drop to increasingly well-organized police work, but declining homicide rates have also occurred in cities with problematic police forces. Reduction in crack cocaine use paralleled the rise and reduction in homicide rates in several American cities, according to the Department of Justice. Jeremy Travis, director of the National Institute of Justice, as quoted by Fox Butterfield, said the Department of Justice tracked homicide rates in six U.S. cities from 1987 to 1993. The report, commissioned by Attorney General Janet Reno, found a close link between crack use and homicide rates. Cocaine use, monitored by testing newly arrested criminals, increased and declined as homicide rates increased and declined. Crack cocaine produces something akin to mania and prolonged use causes paranoia. Crack can cause brain damage and strokes. Cocaine alone can thus produce the equivalent of mental illness and neurologic deficit, two of the three main vulnerabilities to violence. A. Blumstein, F. P. Rivera, and R. Rosenfeld, "The Rise and Decline of Homicide—and Why," *Annual Reviews of Public Health* 21 (2000): 505–41. A. Golub and B. D. Johnson, "A Recent Decline in Cocaine Use Among Youthful Arrestees in Manhattan, 1987 through 1993," *American Journal of Public Health* 84 (1994): 1250–54.

5. *Wrath: Repression and Release—The Effects of Frontal Lobotomy*

1. This case, using a different name, was reported by Dr. Dorothy O. Lewis in her book *Guilty by Reason of Insanity* (New York: Ballantine, 1998).
2. So impressive was this case that the skull of Mr. Gage has been preserved in the medical museum at Harvard University and the damage to his frontal lobes was recalculated with the aid of a computer fairly recently, the results published in *Science*. H. Damasio, T. Grabowski, R. Frank, et al., "The Return of Phineas Gage: Clues about the Brain from the Skull of a Famous Patient," *Science* 264 (1994): 1102–5.

3. The language attributed to Donovan's mother is derived from my notes and memory of the case.
4. The diagnosis of APD has been criticized for its multiple weaknesses. These include shifting diagnostic criteria, absence of symptom weighing, temporal instability, equivalence of some symptoms with substance abuse disorders, inattention of issues to social context, trauma history, and symptom pervasiveness. Neither objective nor projective personality testing reliably differentiates ADP. The diagnosis of ADP does not always indicate criminal activity, let alone incorrigible criminal activity. M. O. Cunningham and T. J. Reidy, "Antisocial Personality Disorder and Psychopathy: Diagnostic Dilemmas in Classifying Patterns of Antisocial Behavior in Sentencing Evaluations," *Behavioral Science and the Law* 16 (1998): 333–51.
5. The Babinski sign, one of the most unequivocal in clinical neurology, refers to the abnormal response of the large toe when the sole of the foot is stimulated with a key. Normally, the toe reflexively bends downward, but if it goes up, it indicates that damage has been done to the corresponding corticospinal tract. This pathway, the longest in the central nervous system, extends from the brain to the spinal cord.
6. These are the most sensitive neuropsychological tools for assessing executive function—that is, judgment, attention, and planning. While it is a bit oversimplified to identify executive function only with the frontal lobes, this region of the brain is important for mediating executive function. Damage to subcortical centers and other regions that interconnect with the frontal lobes can produce similar deficits in executive function to those produced by lesions that are limited to the frontal lobes.
7. Each element of the diagnostic process can stand alone. If the clinical tools of history and physical examination are negative and the MRI shows a brain tumor, the patient has a brain tumor. Similarly, if the history and physical examination indicate brain dysfunction and the MRI is normal, the patient has brain dysfunction. The same is true of neuropsychological tests and the EEG. It is the abnormal finding that is definitive, not the normal one.
8. K. R. Mahaffey, J. L. Annest, J. Roberts, et al., "National Estimates of Blood Lead Levels: United States 1976–1980: Association with Selected Demographic and Socioeconomic Factors." *New England Journal of Medicine* 307 (1982): 573–79.

6. *Immaturity, Mania, Mistreatment, and Miscreancy*

1. A. E. Doyle, J. Biederman, L. J. Seidman, et al., "Diagnostic Efficiency of Neuropsychological Testing with and without Attention Deficit Hyperactivity Disorder," *Journal Consult Clinical Psychology* 68 (2000): 477–88. F. X. Castellanos, J. N. Giedd, P. Eckburg, et al., "Quantitative Morphology of the Caudate in ADHD," *American Journal of Psychiatry* 151 (1994):

665–69. L. Baving, M. Laucht, and M. H. Schmidt, "Atypical Frontal Brain Activation in ADHD," *Journal of the American Academy of Child and Adolescent Psychiatry* 38 (1999): 1363–71. K. G. Sieg, G. R. Gaffney, D. F. Preston, et al., "SPECT Abnormalities in ADHD," *Clinical Nuclear Medicine* 20 (1995): 55–60. H. C. Lou, L. Henrikson, P. Bruhn, et al., "Striatal Dysfunction in Attention Deficit and Hyperactivity Disorder," *Archives of Neurology* 45 (1989): 48–52. SPECT (Single Positive Emission Computerized Tomography) and PET (positron emission tomography) use a radioisotope to measure cerebral circulation. Because increased cellular metabolism demands increased blood flow, PET scans provide an indirect measure of cerebral metabolism by measuring the activity of specific parts of the brain.

2. E. L. Hart, B. B. Lahey, R. Loeber, et al., "Developmental Change in Attention Deficit Hyperactivity Disorder in Boys: A Four-Year Longitudinal Study," *Journal of Abnormal Child Psychology* 23 (1995): 729–49. J. Biederman, S. Faraone, S. Milberger, et al., "Predictors of Persistence and Remission of ADHD into Adolescence: Results from a Four-Year Prospective Follow-up Study," *Journal of the American Academy of Child and Adolescent Psychiatry* 35 (1996): 343–51.
3. Manic-depressive illness, now called bipolar affective disorder, is a classic mental illness characterized by a spontaneous, involuntary alteration in mood. During periods of mania patients are often excited, move around excessively and uncharacteristically, and have racing thoughts and difficulty concentrating, among other symptoms.
4. J. Wozniak et al., "Mania-like Symptoms Suggestive of Childhood-Onset Bipolar Disorder in Clinically Referred Children," *Journal of American Academy Child and Adolescent Psychiatry* 34 (1995): 867–76.
5. Antidepressants would be appropriate for depression, but if they exacerbate mania, mood stabilizers like valproic acid (Depakote), lithium (Lithabid), carbamazepine (Tegretol), gabapentin (Neurontin), and lamotrigine (Lamictal) might be more appropriate, as these control mania. Antipsychotics like quetiapine (Seroquel), haloperidol (Haldol), respiridone (Risperidol), and thioridizine (Mellaril) would be best for schizophrenics.

8. *Nature, Nurture, and Neurology*

1. J. Bouchard, Jr., "Genetic and Environmental Influences on Adult Intelligence and Special Mental Abilities," *Human Biology* 70 (1998): 257–79. G. E. McClearn, B. Johansson, S. Berg, et al., "Substantial Genetic Influence on Cognitive Abilities in Twins 80 or More Years Old," *Science* 276 (2000): 1560–63.
2. T. V. Weisel and D. H. Hubel, "Extent of Recovery from the Effects of Visual Deprivation in Kittens," *Journal of Neurophysiology* 28 (1965): 1060. G. W. Huntley, "Differential Effects of Abnormal Tactile Experience on

Shaping Representation Patterns in Developing Adult Motor Cortex," *Journal of Neuroscience* 17 (1992): 9220.

3. Sally Provence, *Infants in Institutions: A Comparison of Their Development During the First Year of Life with Family Reared Children* (New York: International Universities Press, 1967). R. Spitz and W. Godfrey Cobliner, *First Year of Life: A Psychoanalytic Study of Normal and Deviant Development of Objective Relations* (New York: International Universities Press, 1966). R. Spitz, *No and Yes: On the Genesis of Human Communication* (New York: International Universities Press, 1966).
4. S. Ewing-Cobbs, L. Kramer, M. Prasad, et al., "Neuroimaging, Physical and Developmental Findings After Inflicted and Non-Inflicted Traumatic Brain Injury in Young Children," *Pediatrics* 102 (1998): 300–7. L. M. Schrott, "Effect of Training and Environment on Brain Morphology and Behavior," *Acta Paediatrica. Supplement* 422 (1997): 45–47.
5. T. G. O'Connor and M. Rutter, "The Effects of Global Severe Privation on Cognitive Competence: Extension and Longitudinal English and Romanian Adoptees Follow-up Study Team," *Child Development* 71 (2000): 376–90.
6. H. F. Harlow, M. K. Harlow, and S. J. Suomi, "From Thought to Therapy: Lessons from a Private Laboratory," *American Scientist* 59 (1971): 538–49.
7. C. George and M. Main, "Social Interactions of Young Children: Approach, Avoidance, and Aggression," *Child Development* 50 (1979): 306–18. S. Salzinger, R. S. Feldman, and M. Hammer, "The Effects of Physical Abuse on Children's Social Relationships," *Child Development* 64 (1993): 169–87.
8. C. S. Widom, "Antisocial Personality Disorder in Abused and Neglected Children Grown Up," *American Journal of Psychiatry* 151 (1994): 670–74. J. Paris, H. Zweig-Frank, and J. Gudzer, "Risk Factors for Borderline Personality in Male Outpatients," *Journal of Neurological and Mental Disorders* 182 (1994): 375–80. J. Paris, "Psychological Risk Factors for Borderline Personality Disorder in Female Patients," *Comprehensive Psychiatry* 35 (1994): 375–80. C. L. Harden, "Pseudoseizures and Dissociative Disorders: A Common Mechanism," *Seizure* 6 (1997): 151–55.
9. M. A. Strauss and G. K. Kantor, "Corporal Punishment of Adolescents by Parents: A Risk Factor in the Epidemiology of Depression, Suicide, Alcohol Abuse, Child Abuse, and Wife Beating," *Adolescence* 29 (1994): 543–61.

10. Did You Ever Hear a Baby Cry?

1. H. T. Chugani, "A Critical Period of Brain Development," *Preventive Medicine* 27 (1998): 184–90. P. R. Huttenlocher and A. S. Dabholkar, "Regional Differences in Synaptogenesis in Human Cerebral Cortex," *Journal of Comparative Neurology* 387 (1997): 167–75. L. C. Katz and A. Shatz, "Synaptic Activity and the Construction of Cortical Circuits," *Science* 274

(1996): 1133–35. G. W. Huntley, "Differential Effects of Abnormal Tactile Experience on Shaping Representation Patterns in Developing Adult Motor Cortex," *Journal of Neuroscience* 17 (1997): 9220. F. Sengpiel and C. Blakemore, "The Neural Basis of Suppression and Amblyopia in Strabismus," *Eye* 10 (1996): 250–55.

2. S. Provence, *Infants in Institutions: A Comparison of Their Development During the First Year of Life with Family Reared Children* (New York: International Universities Press, 1967). R. Spitz and W. Godfrey Cobliner, *First Year of Life: A Psychoanalytic Study of Normal and Deviant Development of Objective Relations* (New York: International Universitites Press, 1966). H. F. Harlow, M. K. Harlow, and S. J. Suomi, "From Thought to Therapy: Lessons from a Private Laboratory," *American Scientist* 59 (1971): 538–49.
3. T. G. O'Connor and M. Rutter, "The Effects of Global Severe Privation on Cognitive Competence: Extension and Longitudinal English and Romanian Adoptees Follow-up Study Team," *Child Development* 71 (2000): 376–90. M. Rutter, "Developmental Catch-up and Deficit Following Adoption After Severe, Global, Early Privation. English and Romanian Adoptees Study Team," *Journal of Child Psychology and Psychiatry* 39 (1998): 405–11.
4. D. O. Lewis, C. A. Yeager, Y. Swica, J. H. Pincus, and M. Lewis, "Objective Documentation of Child Abuse and Dissociation in 12 Murderers with Dissociative Identity Disorder," *American Journal of Psychiatry* 154, no. 12 (Dec. 1997): 1703–10.

11. Hitler and Hatred

1. D. J. Goldhagen, *Hitler's Willing Executioners: Ordinary Germans and the Holocaust* (New York: Knopf, 1996).
2. F. McCourt, *Angela's Ashes* (New York: Scribner 1996).
3. The quotations attributed to Trent Scaggs are derived from my memory of the interview.
4. Walter C. Langer, *The Mind of Adolf Hitler: The Secret Wartime Report* (New York: Basic Books, 1972).
5. Adolf Hitler, *Mein Kampf*, English translation (New York: Stackpole, 1939).
6. William Shirer, *Berlin Diary* (New York: Knopf, 1941), p. 137.
7. Quoted by Esther Hecht in the *Jerusalem Post International Edition*, August 10, 1996, pp. 18–19.
8. R. S. Ezekiel, *The Racist Mind* (New York: Viking, 1995).
9. M. A. Strauss and G. K. Kantor, "Corporal Punishment of Adolescents by Parents: A Risk Factor in the Epidemiology of Depression, Suicide, Alcohol Abuse, Child Abuse, and Wife Beating," *Adolescence* 29 (1994): 543–61.

12. Prevention and Treatment

1. G. Breakey and B. Pratt, "Healthy Growth for Hawaii's Healthy Start: Toward a Systematic Statewide Approach to the Prevention of Child Abuse and Neglect," *Zero to Three* 11 (1991): 16–22.
2. S. Murphy, B. Orkow, and R. M. Nicola, "Prenatal Prediction of Child Abuse and Neglect: A Prospective Study," *Child Abuse and Neglect* 9 (1985): 225–35.
3. Elliott Currie, *Crime and Punishment in America* (New York: Metropolitan Books, 1998). D. L. Olds et al., "Long-term Effects of Home Visitation on Maternal Life Course and Child Abuse and Neglect: Fifteen-Year Follow-up of a Randomized Trial," *Journal of the American Medical Association* 278 (1997): 637–43. J. B. Hardy and R. Streett, "Family Support and Parenting Education in the Home: An Effective Extension of Clinic-Based Preventive Health Care Services for Poor Children," *Journal of Pediatrics* 115 (1989): 927–31. R. B. Earle, "Helping to Prevent Child Abuse and Future Consequences: Hawaii Healthy Start," National Institute of Justice, Washington, D.C., 1995.
4. L. Schweinhart, H. V. Barnes, and D. Weikart, "Significant Benefits: The High/Scope Perry Preschool Study Through Age 27, Ypsilanti, Michigan" (High/Scope Press, 1993). D. S. Gomby et al., "Long-term Outcomes of Early Childhood Programs: Analysis and Recommendations," *The Future of Children* 5, no. 3 (winter 1995).
5. V. Seitz, "Intervention Programs for Impoverished Children: A Comparison of Educational and Family Support Models," *Annals of Child Development* 7 (1990): 84–87. J. R. Lally, P. L. Mangione, and A. S. Honig, "The Syracuse University Family Development Research Program," in Douglas R. Powell, ed., *Annals of Advances in Applied Developmental Psychology* 3 (1988): 79–104.
6. T. Haspel, "Betablockers and the Treatment of Aggression," *Harvard Review of Psychiatry* 2 (1995): 274–81.
7. S. Mitchison, K. J. Rix, E. B. Renvoize, et al., "Recorded Psychiatric Morbidity in a Large Prison for Remanded and Sentenced Prisoners," *Medical Science and the Law* 34 (1994): 324–30.
8. M. H. Sheard, J. L. Marini, C. I. Bridges, et al., "The Effect of Lithium on Impulsive Aggressive Behavior in Men," *American Journal of Psychiatry* 13 (1976): 1409–13.
9. H. G. Kennedy, R. C. Iveson, and O. Hill, "Violence, Homicide, and Suicide: Strong Correlation," *British Journal of Psychiatry* 175 (1999): 462–66.
10. Bupropion (Welbutrin) and Venlafaxine (Effexor) are newer antidepressants that many psychiatrists believe are less likely than other antidepressants to precipitate mania in bipolar disorder. The factual basis for this widely held belief has not been established.

11. J. A. Yesavage, "Correlates of Dangerous Behavior by Schizophrenics in a Hospital," *Journal of Psychiatric Research* 18 (1984): 225–31.
12. P. J. Taylor and J. Gunn, "Violence and Psychosis: I. Risk of Violence Among Psychotic Men," *British Medical Journal* 288 (1984): 1945–49.
13. J. R. Stevens, and B. P. Hermann,"Temporal Lobe Epilepsy, Psychopathology and Violence: The State of Evidence," *Neurology* 31 (1981): 1127–32.

Appendix: Tools of Diagnosis: History, Physical Examination, and the Role of Tests

1. The frontal lobes have been divided into three overlapping anatomical-functional regions: (1) the cingulate gyrus is a C-shaped band of gray matter that runs from the front to the back of the brain in the middle. Lesions of the cingulate gyrus produce apathy. (2) The dorsolateral region is underneath the temples and subserves executive functioning. Neuropsychological tests and the physical examination are good tests of this region. (3) The orbital frontal lobe is behind the forehead and above the eyes. It governs personality. It was this region that was destroyed in the case of Phineas Gage. Neuropsychological tests and the physical examination are often normal, even when this region is severly disrupted and behavior extremely abnormal. Each of the three regions connects to the basal ganglia and the thalamus. These are masses of nerve cells that are grouped together deep in the brain, outside the frontal lobe. Lesions of the basal ganglia and thalamus can produce frontal-type deficits in cognition and behavior. One of the conditions that is especially difficult to differentiate from frontal lobe dysfunction is antisocial personality disorder. Personality inventories like the Minnesota Multiphasic Personality Inventory (MMPI) cannot differentiate personality disorders from frontal damage. Raine and his colleagues have suggested that antisocial personality disorder is, in fact, the result of frontal lobe dysfunction. A. Raine, T. Lencz, S. Bihrle, et al., "Reduced Prefrontal Gray Matter Volume and Reduced Autonomic Activity in Antisocial Personality Disorder," *Archives of General Psychiatry* 57 (2000): 119–27. A. Raine, M. Buchsbaum, and L. LaCasse, "Brain Abnormalities in Murderers Indicated by Positron Emission Tomography," *Biological Psychiatry* 42 (1997): 495–508.
2. L. K. Jenkyn, D. B. Walsh, and A. G. Reeves, "Clinical Signs in Diffuse Cerebral Dysfunction," *Journal of Neurology, Neurosurgery, and Psychiatry* 40: (1977) 956–66. L. R. Jenkyn, A. G. Reeves, W. T. Whiting et al., "Neurologic Signs in Senescence," *Archives of Neurology* 42 (1985): 1154–57.
3. M. D. Lezak, *Verbal Functions in Neuropsychological Assessment* (New York and Oxford: Chiverst Press, 1976), pp. 265–69.
4. A. R. Luria, *Higher Cortical Functions in Man,* 2nd ed. (New York: Basic Books, 1962).
5. Ibid.

6. E. A. Rodin, "Impaired Ocular Pursuit Movements," *Archives of Neurology* 10 (1964): 327–31.
7. G. Paulson and G. Gottlieb, "Developmental Reflexes: The Reappearance of Foetal and Neonatal Reflexes in Aged Patients," *Brain* 91 (1968): 37–52.
8. L. R. Jenkyn et al., "Clinical Signs in Diffuse Cerebral Dysfunction," *Journal of Neurology, Neurosurgery, and Psychiatry* 40 (1977): 956–66; C. Bae et al., "Neurologic Signs Predict Periventricular White Matter Lesions on MRI," *Neurology* 50 (1998): A448.
9. J. Grafman et al., "Frontal Lobe Injuries, Violence, and Aggression: A Report of the Vietnam Head Injury Study," *Neurology* 46 (1996): 1231–38.
10. P. Y. Blake, J. H. Pincus, C. Buckner, "Neurologic Abnormalities in Murderers," *Neurology* 45 (1995): 1641–47. I have examined about 150 murderers, most of them on death row, in the past twenty-five years.
11. M. Nuwer, "Assessment of Digital EEG, Quantitative EEG, and EEG Brain Mapping: Report of the American Academy of Neurology and the Clinical Neurophysiology Society," *Neurology* 49 (1997): 277.
12. A. Raine et al., op. cit., 2000.
13. S. E. Shaywitz, B. A. Shaywitz, K. R. Pugh, et al.,"Functional Disruptions of the Organization of the Brain for Reading in Dyslexia," *Proceedings of the National Academy of Science* 95 (1998): 2636–41.